THE POWER
SUPPLY HANDBOOK

THE POWER SUPPLY HANDBOOK

BY THE EDITORS OF 73 MAGAZINE

TAB BOOKS

BLUE RIDGE SUMMIT, PA. 17214

FIRST EDITION

FIRST PRINTING—JANUARY 1979

Copyright © 1979 by TAB BOOKS

Printed in the United States of America

Library of Congress Cataloging in Publication Data

Main entry under title:

   The Power supply handbook.

   Includes index.
      1. Electronic apparatus and appliances—Power supply. I. 73 magazine for radio amateurs.
TK7868.P6P68   621.381'04'4   78-31272
ISBN 0-8306-9806-X
ISBN 0-8306-8806-4 pbk.

Foreword

All too often the power supply takes a backseat to other electronic circuits. It is a simple matter to say "...and build a power supply to furnish the required voltage and current." But when it comes down to actually building the power supply, it isn't as easy as you might imagine without some sort of guidance, unless you're an electronic engineer with a warehouse of parts at your disposal.

Well, the guidance is here in these 420 pages of *The Handbook of Power Supplies*. This is a collection of every conceivable type of power supply for the electronic hobbyist, ham, experimenter, engineer, technician, CBer, and do-it-yourselfer. Be it a lab supply, inverter, or special purpose type, you can build it, or even possibly design it from scratch, with the information compiled here.

With the onset of home computers in kit form and the like, this book takes on even more significance. These mini-brains require power supplies of special caliber to protect the microprocessors and other logic circuits from damage. Many of these aspects of power supply selection are covered in Chapter 1.

Regulated low-voltage power supplies are by far the most often used in today's solid-state circuitry. Chapter 2 covers a multitude of this type of supply. You'll find regulated supplies with rated outputs from 0.35V on up to 50V in this chapter.

There's going to be a case when you find a power supply that fits your requirements in every way except one minor detail, and that is the purpose of Chapter 3. This chapter is a collection of mostly add-on gadgets that can be attached to an existing power supply, a voltage splitter for example.

Although it seems as though vacuum tubes have almost disappeared, there's still a lot of equipment floating around that rely on them. And these pieces of equipment also require power supplies, but they need higher voltages than our common regulated low-voltage supplies of Chapter 2 deliver. Chapter 4 covers a few of these high-voltage supplies chiefly for transmitters and the like. One of these babies delivers 3000V!

Chapters 5, 6, and 7 cover inverters, AC-to-DC converters, and DC-to-DC converters, respectively. Battery chargers, another type of power supply, are the subject of Chapter 8, including special types for silver-zinc and gelled cells. Sensing circuits, both voltage and current, are covered in Chapter 9. Mobile power and accessories are handled in Chapter 10.

A final chapter deals with special purpose AC supplies. Whatever the application, you will have no trouble finding or designing a power supply with the wealth of knowhow contained on the following pages.

Mike Fair
TAB Editorial Staff

Contents

Chapter 1
Power Supply Facts

This chapter contains all of the raw data that you need to engineer your own power supply. You'll probably discover that in many cases some of the power supplies illustrated in later chapters will come very close to what you require for a particular circuit; however, there are those slight variations that you need to know how to handle. This is the purpose of the information included here—to smooth out those ripples.

You'll also read about a lot of hard-core theory in this chapter. There are several sections that deal with actual power supply design from scratch. Then there are sections that deal with power transformer ratings, how to build transformerless supplies, how to make the most of triacs, and even how to test a supply for home computer applications.

DESIGNING A REGULATED SUPPLY

Regulated power supplies are a mystery. Almost every IC construction project includes a regulated supply and most solid-state equipment built for 117V also has a regulated supply or supplies to power the low-voltage solid-state devices. But the mystery is that while most hobbyists have a good idea of how to use transistors and integrated circuits in simple applications, few have the remotest idea of how the regulated

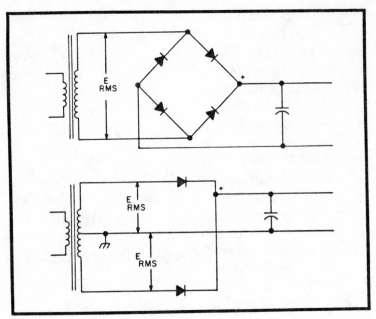

Fig. 1-1. Schematic of full-wave rectifier circuits. (A) Bridge. (B) Center-tapped transformer.

supply works and fewer still could design one from scratch if required to.

This portion of the chapter deals with the operation of regulated power supplies for low-voltage applications and gives all the necessary information needed to design a regulated supply from scratch. Information is given so that the designer may select the proper components such as the transformer, diodes, capacitor, regulator, pass transistors and heat sinks. Sources for all parts are given so that the designer won't be stuck for some hard to find parts.

The regulated power supply consists of an unregulated DC power supply feeding a regulating circuit. The unregulated DC supply may consist of a full-wave rectifier feeding a filter capacitor as shown in Fig. 1-1 or it may be a battery used in a mobile or portable installation. The regulating circuit may be a circuit made up of discrete components or it may be a regulating IC, such as the NE550. Components and design options are chosen according to the voltage and current requirements of the project needing the regulated supply.

Integrated circuit voltage regulators are commonly used today, rather than discrete components, because of their low cost and ease of use. The basic design comes from the Signetics *Digital, Linear, Mos* manual and is based on the Signetics NE550 regulator IC. This basic design is simple and permits numerous output voltages and limiting currents by merely selecting readily available resistor combinations.

The DC Power Supply

The DC power supply used for most low-voltage power supplies is a capacitive load circuit as shown in Fig. 1-1. Inductive filters are occasionally used instead of capacitors, but high-value, high-current inductors are more difficult to locate and more expensive than low-voltage high-value capacitors. Either a full-wave (Fig 1-1) or half-wave (Fig. 1-2) circuit may be used to supply the DC; however, a full-wave circuit is preferred because it provides better basic regulation. The full-wave circuit is used in this design.

In order to determine the voltage and current ratings of the components to be used in the unregulated DC power supply, it is first necessary to determine the voltage and current requirements of the equipment or device to be powered. When determining these power requirements it is best to allow reasonable safety factors in order to prevent overheating and to insure that the equipment will operate correctly. Normally a current safety factor of 10% is allowed in cases

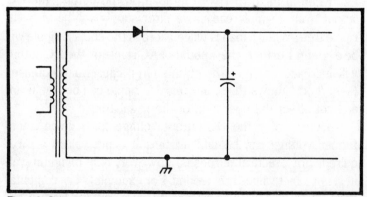

Fig. 1-2. Schematic of half-wave rectifier circuit.

where peak current is being drawn 50% of the time or less. In all other cases allow a safety factor of one-third.

To determine the required current rating for the transformer use the formula $I = 1.3 \times I_p$, where I_p is the anticipated peak current requirements of the equipment.

The designer may design the basic unregulated DC power supply so that the DC output voltage is anywhere from 30% to 98% of the peak AC voltage of the transformer. If a large value filter capacitor is used, the 98% value may be achieved and little ripple will appear on the output of the basic supply. Unfortunately, very high-capacitance capacitors are expensive and in some cases may be hard to find. Smaller value capacitors are less expensive and easy to locate but will give lower DC outputs and will produce appreciable ripple on the output. For given DC output voltage (under load), the AC output of the transformer will have to be higher for small filter capacitors as compared to large value filter capacitors. Note that in general it is less expensive to use a transformer of higher voltage with a low-capacitance capacitor for a given DC output than it is to use a large capacitor and a lower voltage transformer to produce the same DC voltage. This is logical since the cost of a transformer does not increase appreciably as the voltage goes up, while the cost of a capacitor increases significantly as the value of capacitance goes up. In order to minimize the cost of the supply, one of the design factors is to keep the filter capacitor to moderate size and low cost.

Keep in mind that with no load on the output, the DC output from a simple capacitive filter supply will be virtually ripple free. When a load is placed across the supply, ripple will be evident. Further, the amount of AC ripple on the DC output will increase as the size of the filter capacitor decreases (everybody knows that), but this AC ripple can be significant and not affect the operation of the regulator.

Determining the DC output voltage for a given transformer voltage can be difficult task if exact values are required. For practical purposes, however, only minimum values, not exact values are needed. For example, if our computations show us that we will get 18V DC output from a DC supply, but we really get more than 18V, then this is of no consequence. We only want to assure ourselves that we will get at least the minimum required under load. With this in mind, the following formulas can be used to determine the AC (RMS) value of the transformer required:

$$E_{PEAK} = 1.4 \times E_{MS}$$
$$E_{OUT} = .71 \times E_{PEAK}$$

thus,

$$E_{OUT} = 0.71 \times 1.4 \times E_{RMS}$$

where E_{OUT} is the minimum DC output voltage from the unregulated DC power supply, and E_{RMS} is the secondary voltage of the transformer. In summary, the anticipated DC output voltage under load from a simple unregulated supply as shown in Fig 1-1 will be equal to or greater than the AC voltage from the secondary of the transformer. This will only hold true if the current ratings of the transformer are not exceeded. The above formula takes into consideration that a moderate size capacitor will be used and is based on the assumption that ripple on the DC output voltage can be 10% or less. The NE550 regulator IC, which is used for the design presented here, can tolerate 10% ripple provided that the lowest DC input voltage (low part of the ripple) is at least 3V higher than the desired DC regulated voltage. Thus, we will have to consider the DC input voltage to be the bottom of the ripple as shown in Fig. 1-3. Note that the peak voltage cannot be higher than the maximum ratings of the NE550. As defined, $E_{DC\ INPUT} = E_{REG} + 3$, where E_{REG} is the desired regulated voltage.

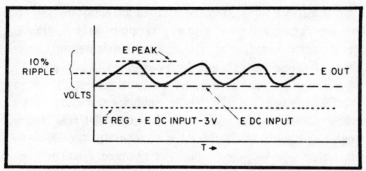

Fig. 1-3. Allowable ripple.

Current	Voltage	Distributor	Part #	Approx. Cost
3 A	12.6 V	Radio Shack	273-1511	$4.29
2 A	25.2 V	Radio Shack	273-1512	4.29
6.6 A	6.3 V	Verada		3.75 ea., 3/$10
2.5 A	30 V	M. Weinshenker		4.85
12 A	18 V	VHF Engineering		

Fig. 1-4. Listing of possible transformers to use in designing a regulated power supply.

The DC input voltage is also 95% of E_{OUT} (E_{DC} input is 5% lower then E_{OUT} because of ripple); thus $E_{DC\ INPUT} = 0.95\ E_{OUT} = 0.95\ E_{RMS}$. Solving the two equations gives $E_{RMS} = (E_{REG} + 3)/0.95$. We now have a very simple formula to use to determine the secondary (E_{RMS}) value of the transformer given only what we want for a regulated voltage and assuming that we will not exceed the manufacturer's current rating for the transformer chosen.

These formulas given for I and E_{RMS} will hold true for virtually all low-voltage, high-current supplies provided that good quality properly designed transformers are selected. The transformers recommended mentioned later fall into that class. If low-grade transformers with high internal resistance are used, then E_{RMS} may approach a value of $(E_{REG} + 3)/0.5$.

As an example, assume that we want a power supply to deliver 5A at 12V regulated. The minimum ratings of the transformer would be determined as follows:

$$I = 1.3 \times I_p = 1.3 \times 5 = 6.5A$$

$E_{RMS} = (12 + 3)/0.95 = 16V$. Looking through the various catalogs you probably won't find a transformer that has a secondary exactly matching our requirements, but you would find one that exceeds the requirements. Looking at Fig. 1-4 we find that Verada in Lowell, Massachusetts, is offering $3.75 each or 3 for $10. Three of these transformers with their secondaries in series, primaries in parallel, (Fig. 1-5) will give an output of 18.9V at 6.6A at a cost of $10.00. This is well less than the equivalent single transformer would cost if purchased from an electronic supply house. Thus three Verada transformers are used in this design with an RMS secondary voltage of 18.9V.

It is a good idea to check the peak DC output voltage obtainable under any circumstances to see that this voltage does not exceed the voltage ratings of the NE550 regulator. The maximum voltage is given by $E_{MA} = 1.4 \times E_{RMS}$. Thus in our case $E_{MAX} = 1.4 \times 18.9 = 26.5$V. The maximum voltage rating of the NE550 is 40V. We are within the limits in this case. In a case where E_{REG} is 37V, the maximum allowable for the NE550, or any case where E_{MAX} exceeds 40V, then the circuit in Fig. 1-6 must be used to provide the DC input voltage to the regulator.

Selection of diodes can be made in a fashion similar to the transformer. Diodes in a full-wave configuration pass only one-half the total current, thus $I_D = 0.5I$, where I_D represents the current requirements of the diodes. In our example, the maximum current is 6.5A, so the diodes would have to handle 3.25A each. Since this an oddball value, the next higher current rating would be used such as diodes with a 6A rating. To be conservative for low-voltage supplies, the voltage ratings of the diodes should be greater than the maximum peak voltage that can be encountered. For a bridge rectifier config-

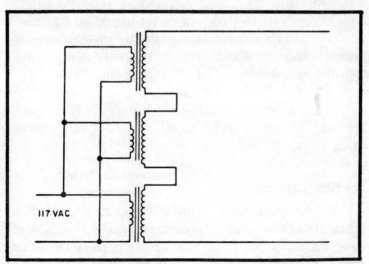

Fig. 1-5. Schematic of three transformers with their secondaries wired in series and primaries in parallel.

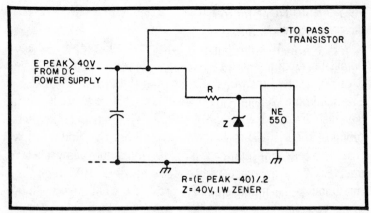

Fig. 1-6. Schematic of regulator protective circuit when EPEAK is greater than 40V.

uration, each diode should have a PIV (peak inverse voltage) rating of four times the E RMS value of the transformer secondary, while for a center-tapped rectifier configuration the PIV should be six times the E RMS value of the transformer secondary. In our example, diodes in a bridge would have to have a minimum PIV rating of 75.6V, while diodes in the half-wave configurations should have a rating of 113V. These are oddball values, so we would use diodes of the next higher rating. A bridge with 100 PIV rating could be used (Poly Paks 10A, 100V bridge) or two diodes in center-tapped configuration with a rating of 150V would do. Note that the current rating for a complete diode bridge (as compared to individual diodes in a bridge) is not divided in half. In this example 6.5A is the requirement, so a 10A bridge would be required. Diode bridges are preferred since they are usually epoxy encapsulated and may be mounted directly to a heat sink without having to worry about mica insulators and special means to provide insulation.

The Filter Capacitor

The filter capacitor smooths the pulsating DC and gives steady state DC with some percentage of ripple on top. One of design criteria is that we can tolerate 10% ripple. If the wrong capacitor is chosen, the ripple may exceed 10% (if the capacitor is too small) and the output voltage may be too low, causing the regulator to regulate poorly for heavy loads. If the capacitor is too large, the ripple will be smaller and the output voltage from the unregulated supply will be higher, but this is of little consequence to us. Thus, we have to determine the minimum size of the capacitor. Note that excessively large filter capacitors can cause enormously large surge currents through the diodes during turn on. Most silicon diodes can handle large surges for an instant or two so this shouldn't be too much of a problem. If the designer sticks close (50% to 100% larger) to the value of capacitance determined during calculators, problems should not be encountered with popping diodes.

To determine the proper capacitor, it is necessary to first determine the load resistance. This load resistance is determined by the formula $R_L = E_{OUT}/I$, where E_{OUT} is the output voltage we need from the unregulated DC supply in order to supply E_{REG}. I is the maximum current to be drawn. Note that load includes power dissipated in the regulating circuitry. In our example $E_{OUT} = E_{RMS} = 18.9V$ (theoretical E_{RMS} was 16V, but we use three transformers to give us 18.9V) and I was 6.5A. Thus $R_L = 18.9/6.5 = 2.9\Omega$. The basic formula to be used for the value of the capacitor is $2\pi f R_L C = 5$, where $\pi = 3.14$ and $f = $ the line frequency. C is the desired capacitance in farads. Solving the equation for C we get $C = 5/(6.28$ $f R_L)$. In our example the line frequency is 60 Hz and $R_L = 2.9\Omega$. Thus $C = 5/(6.28 \times 60 \times 2.9) = 0.0046$ or 4,600 μF. Since 4,6000 μF is not a stock value, the next higher value would be used. The voltage rating of the capacitor should be at least double the E_{RMS} voltage which in our case is 38V.

Some purists may question the 10% tolerable ripple figure previously given. This is understandable since for tube-type power supplies no ripple was tolerable. While this 10% figure may seem like a lot, remember that the only thing that the 550 regulator cares about is that the lowest DC input voltage is at least 3V above the regulated output voltage. The amount of ripple on top of this minimum DC input voltage is insignificant as long as maximum ratings are not exceeded.

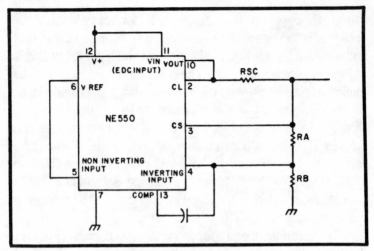

Fig. 1-7. Schematic of NE550 regulator circuit.

The Regulator

The NE550 regulator is an operational amplifier with internal reference voltage and current limiting capabilities. The operational amplifier compares an internal standard reference voltage (internal zener) fed into the noninverting input with a sample of the desired regulated voltage. The difference between the standard voltage and a sample of the regulated output is amplified and inverted, producing a control voltage. This control voltage controls a pass transistor which is in series with the regulated output. As the regulated output drops, the control voltage increases, which in turn causes the regulated output to increase. A stable point is eventually reached where the output voltage remains constant. This stable point depends on the ratio of two resistors (R_A and R_B in Fig. 1-7) connected as a voltage divider to deliver a sample of the regulated output to the inverting input of the operational amplifier. By changing the ratio of the values of the two resistors, the output of the voltage divider changes, which in turn produces a change in the regulated output. The value of the regulated output, thus can be simply changed by altering the ratio of the values of two resistors. Figure 1-8 gives various values of R_A and R_B for selected values of regulated voltage.

| | 1% Resistors | | 5% Resistors | | |
E_{reg}	R_A k Ohms	R_B k Ohms	R_A k Ohms	R_B k Ohms	Trimming Resistor
5	6.13	2.97	5.6	2	1k
10	12.3	2.39	11	2	1k
12	14.7	2.31	13	2	2k
13	16.0	2.29	15	2	2k
15	18.4	2.24	16	2	5k
20	24.5	2.18	22	2	5k
30	36.8	2.11	33	2	5k

Fig. 1-8. Listings of values of resistors R_A and R_B at 1% and 5%.

Current limiting is provided by connecting a current-sensing resistor between the base and emitter of a transistor as shown in Fig. 1-9. When a current through the current-sensing resistor reaches the value of I_{LIMIT}, the value of current at which to limit the current, a voltage of 0.64V appears across the emitter-base terminals of the limiting transistor. This limiting transistor starts to conduct and causes the base voltage on an internal pass transistor to be lowered. Since the base voltage of this internal pass transistor is lowered, the output voltage from the regulator is lowered. The output voltage does not drop to zero when I_{LIMIT} is reached, but is merely lowered the right amount needed in order to maintain the output current at I_{LIMIT}. The output current is not shut off but stays at the value of I_{LIMIT}. When the load resistance increases to a point where the output current

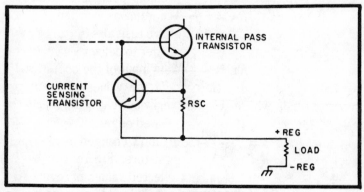

Fig. 1-9. Schematic of current-limiting circuit.

can fall below I_{LIMIT}, then the output voltage goes back to E_{REG}. This is a very important concept to understand, because if I_{LIMIT} is set at or below the current that the load draws, then regulation will be lost. For example, assume that I_{LIMIT} is set at 6.5A and that the regulated supply was designed for use with a two meter SSB transceiver. The output of the regulated supply is 12V and the average current drawn by the SSB transceiver is 4A. An SSB transceiver by its nature can draw very high peaks of current and it is not unreasonable to assume that on current peaks the transceiver might draw 10A of current. On voice peaks with I_{LIMIT} set at 6.5A, the current limiting will prevent the SSB transceiver from drawing more than 6.5A. The only way that it can do this is to reduce the output voltage, destroying regulation. It is possible that on voice peaks only 6V or 7V would be delivered to the transceiver. Low voltage will in most cases adversely affect the operation of any piece of gear and in some cases can cause damage. It is important to understand that *current limiting protects the power supply*, not the equipment. If it is necessary to protect the equipment from low voltage as well as over current, then a detecting circuit must be employed to shut down and latch the voltage off when over current is detected. This type of circuit is another subject and will not be covered here.

To select a given regulated voltage, R_A and R_B must be chosen according to the formulas.

$$R_A = \frac{2000}{1.63} \quad E_{REG}$$

$$R_B = \frac{2000 \, E_{REG}}{(E_{REG} - 1.63)}$$

In our example where we want an output voltage of 12V:

$$R_A = \frac{2000}{1.63} \times 12 = 14,700\Omega$$

$$R_B = \frac{2000 \times 12}{(12 - 1.63)} = 2,310\Omega$$

The value of the current limiting resistor is determined by the equation $R_{SC} = 0.64/I_{LIMIT}$, where I_{LIMIT} is the value of current at which to limit. In the power supply in the example, $I_{LIMIT} = 6.5A$ which is the basic design value of the supply. Thus $R_{SC} = 0.64/6.5 = 0.1\Omega$ (approximately).

From a practical standpoint it is not always possible to obtain precision resistors in the values you want without having to resort to ordering from mail order houses. While adjustment of the regulated output voltage may be achieved by using the circuit in Fig. 1-10 instead of two fixed resistors, the fixed resistors provide convenience and greater reliability over a circuit where a potentiometer may be jarred or misset by inexperienced personnel. (Fixed resistors will also keep twiddlers from increasing the output by increasing the voltage. ZAP!) Five percent tolerance resistors may be used in place of the one percent resistors normally required for precise voltage determination. Figure 1-8 gives the value of 5% resistors to be used for selected voltages. When using 5% resistors, a trimming resistor as shown in Fig. 1-11 should be

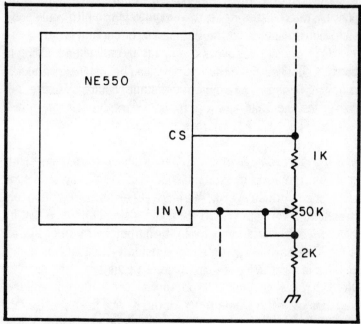

Fig. 1-10. Schematic of variable voltage supply.

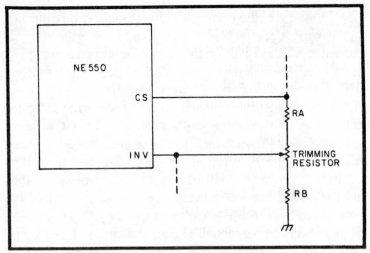

Fig. 1-11. Schematic of variable voltage supply with resistors R_A and R_B at 5%. The trimming resistor should also be 5%.

used. The value of this trimming resistor is also given. Every power supply could be constructed as shown in Fig. 1-10 to provide adjustability over the entire range, but it is safer to limit the adjustment to a fraction of the entire range to prevent possible misadjustment into a voltage region which could possibly cause damage to the equipment being powered.

One percent resistors are handy to use when building a batch of supplies since the resistors can be installed without a trimming resistor and adjustment is not required. Figure 1-8 also gives the values of 1% resistors required for selected voltage. These 1% resistors can be obtained from Allied Electronics.

The formulas given for R_A and R_B have been derived for optimum temperature stability of E_{REG}. With very little sacrifice at all in this regard, the basic regulating circuit may be made into a variable voltage regulating supply. Notice that in Figure 1-8 the variation of R_B is small over the range of 10V to 40V. If we assume that R_B is a constant over that range and that R_A is a variable, we can make a variable voltage supply. Note that we can select a potentiometer to permit voltage variation over the whole range of the device from 2V to 37V, but the temperature stability at voltages of less than 10V will not be as good as the temperature stability in the range above 10V. Figure 1-10 shows the circuit for a variable voltage supply. One unique thing about variable voltage capability is that the current limit stays the same over the entire range. I_{LIMIT} is not dependent on the value of regulated voltage, but is dependent only on the value of R_{SC}.

It is possible to switch different R_{SC}, current sensing resistors in and out of the circuit; however, from a practical standpoint problems arise. Switch contacts have resistance (both in a clean and dirty state). This resistance will become a part of R_{SC} and will affect the value of I_{LIMIT}, making I_{LIMIT} smaller as resistance increases. In addition, the contact resistance may vary from one switch cycle to the next, causing I_{LIMIT} to vary. A potentiometer conceivably could serve the purpose, but potentiometers also get dirty and very low resistance potentiometers are extremely hard to find.

The basic regulator circuit by itself as shown in Fig. 1-7 would not handle the current of 6.5A required in our example. The basic circuit would only handle 150 mA. In order to handle additional current, the circuit in Fig. 1-12 would have to be used. This circuit uses an external pass transistor to control the higher current. This transistor is selected by choosing a

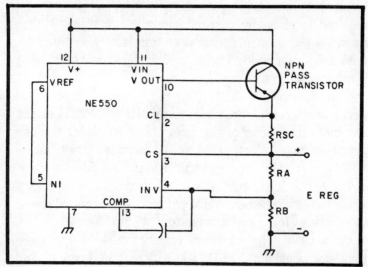

Fig. 1-12. Schematic of the NE550 regulator driving a pass transistor.

Wattage	Max Current	Minimum dc Current Gain (h_{FE})	Max Voltage	Thermal Derating Factor	Transistor #	Motorola HEP Equivalent
20	2 A dc	25	60	.133 W/°C	MJ2249	HEP241
40	2	25	125	.266	2N5050	HEP241
87.5	4	30	40	.5	MJ480	HEP247
85	4	750	60	.343	MJ4200	–
90	10	20	60	.718	MJE3055	S5001
115	15	20	60	.657	2N3055	HEP704
150	20	500	40	1.2	2N6355	–

Fig. 1-13. Listing for transistor selection.

transistor with sufficient current carrying capacity and sufficient DC gain. In our case the current carrying capacity must be 6.5A minimum. The DC gain can be obtained from the formula:

$$\text{Current gain} = \frac{\text{output current}}{0.15}$$

The DC gain can be found in Fig. 1-13 under the heading h_{FE}. Note that h_{FE} is usually specified for a particular current. This specified current may not be the value as the design current of the supply; thus, h_{FE} is really an approximation for most purposes. This approximation should cause few problems if the transistors in Fig. 1-13 are used. In our case the required current carrying capability is 6.5A and the DC gain required is given by current gain = 6.5/0.15 = 43. Our minimum requirements for a pass transistor would then be 6.5A and a gain of 43. The voltage rating should be at least $2 \times E_{RMS}$, or 38V.

In addition to h_{FE} (DC gain) and maximum current, it is necessary to consider the heat dissipation of the device. The maximum dissipation of the pass transistor under working conditions may be estimated by the formula $P_D = (E_{RMS} - E_{REG} \times I_{LIMIT}$. For the example we are considering $P_D = (19 - 12) \times 6.5 = 46W$. It is not unreasonable to use a 25% safety factor for the power dissipation, so the required power dissipation for this transistor would be 46 + 1/3 × 46) = 61W. The minimum requirements for our pass transistor would thus be 61W dissipation, 6.5A, h_{FE} of 43, and maximum voltage of 38V. To give yourself maximum flexibility, a good manual which lists transistor characteristics is handy (but not required). The Motorola *Semiconductor Data Library* is a good guide to go by. If you don't have a book listing transistor characteristics, then you may use Fig. 1-13. Using Fig. 1-13 we find that a 2N6355 is suitable since it exceeds the specifications needed for this application. The 2N6355 has a dissipation of 150W, 20A current rating, h_{FE} of 500, and maximum voltage of 40V. Note that this transistor is a high-gain Darlington transistor and was chosen since it is readily available. It is wise to stay away from transistors with extremely high gains since these transistors can sometimes go into oscillation all by themselves.

The above calculations assume that the supply will handle intermittent shorts. If the supply must handle continuous shorts, then $P_D = R_{RMS}/I_{LIMIT}$. At this point, the power supply design as shown in Fig. 1-12 is complete except for considering the heat sinking requirements.

Heat Sinking

As everyone knows, semiconductors dissipate heat and in some cases get very hot to the touch. But just how hot can semiconductors get without being destroyed? The data sheets for both diodes and transistors give power derating factors for reducing the allowable dissipation or allowable current for a given temperature rise. These factors must be used to prevent overheating and burnout of the device. Note that a power derating factor is a multiplier, based on temperature, which serves to lower the power rating as the temperature of the device increases. Figure 1-13 gives derating factors for selected transistors.

Let us assume that in this power supply we used 100V, 6A diodes such as the Motorola 1N3880. To determine the power dissipation of the diodes use the empirically derived formula $P_{DIODE} = 1 \times 1.5$; thus for our example $P_{DIODE} = 3.25 \times 1.5 = 5W$ (approximately). Furthermore from Fig. 1-14 we find that the maximum usable temperature without derating is 100°C.

Max. Current	Max. Volts	Maximum Temp. Without Derating	Diode #	
1	200	–	HEP 156	Heat Sink Not Required
2.5	1000	–	HEP 170	Heat Sink Not Required
6	100	100°	1N3880	(HEP 153)
6	200	100°	1N3881	(HEP 153)

Fig. 1-14. Listing for diode selection. Temperatures shown are degrees Celsius.

At this point we must determine the cooling capacity of the heat sink needed. Heat sinks are rated in degrees Celsus per watt. In other words, for every watt of dissipation, the temperature of the heat sink will rise so many degrees Celsius above ambient room temperature. If a given heat sink is rated as 10° C/W, then the temperature of the heat sink will rise 10° for every watt dissipated. The formula for determining the cooling capacity of the heat sink needed is:

$$\text{Cooling cap (°C/W)} = \frac{\text{max} - \text{ambient temp}}{\text{watts dissipation}}$$

From Fig. 1-14, the maximum temperature without derating is 100° C. While our actual dissipation will be 5W, a 25% safety factor applied to 5W gives 6.25W or 7W (rounded upward). Using the formula above, Cooling cap = (100 − 23)/7 = 11° C/W. Figure 1-15 gives various heat sinks selected according to various cooling capacities. As can be seen, the Thermalloy #6111B might be used. This heat sink will produce a temperature rise of 10° C/W. When selecting a heat sink keep in mind that the specifications of the heat sink must give *less* heat rise than the device. In other words, the degrees rise per watt will be smaller or equal to the permissible heat rise of the device. Note that when mounting the diodes to the heat sinks that silicone heat conducting grease should be used on the diodes as well as the mica insulating washers as shown in Fig. 1-16.

As previously determined the transistor chosen will dissipate 61W under the conditions defined. In order to determine the amount of heat sinkng required it is necessary to determine the maximum permissible device temperature rise for a dissipation of 61W, the maximum dissipation that we previously calculated. If you have access to charts, this temp-

MFR	#	Cooling Capacity (Thermal Resistance)
Thermalloy	6111B	10°C/W
"	6176B	4
"	6401B	2.3
"	6403B	1.5
"	6421B	1.0
"	6441B	.54
"	6690B	.28

Fig. 1-15. Listing for heat sink selection.

erature rise is easy to find. In the absence of charts, the derating factor given in Fig. 1-13 is used. The derating factor is a factor that tells us the reduction in wattage which must be applied against the maximum wattage as the temperature rises. For example, for the 2N6355 the derating factor is 1.2W/°C. Thus for every degree Celsius above room temperature, the wattage must be reduced by 1.2W. As the temperature rises, the power dissipation decreases. This factor can be used to determine the maximum permissible device temperature rise. This temperature rise is determined by the formula:

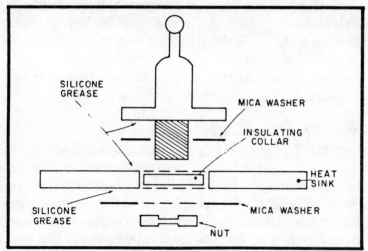

Fig. 1-16. Method of mounting diodes to heat sink.

$$\text{Temp rise} = \frac{(\text{maximum wattage} - \text{required wattage})}{\text{derating factor}}$$

Thus in our example, the maximum wattage is 150W, the required wattage dissipation is 61W, the derating factor is 1.2 W/°C, and

$$\text{Temp rise} = \frac{150 - 61}{1.2} = 74°$$

In this case, as long as the temperature of the device does not increase by more than 74°C, we will be operating within safe bounds. Note that we are talking about a temperature rise in this case, so in order to use the previous formula we would have to add the ambient room temperature to get the maximum temperature, but then we turn right around and subtract the room temperature. Thus, the previous formula can be modified for cases where we are talking about temperature rise as follows:

$$\text{Cooling capacity} = \frac{\text{temperature rise}}{\text{dissipation (watts)}}$$

In our example we have: Cooling capacity = 74/61 = 1.2° C/W. The heat sink required must have a thermal resistance of less than 1.2° C/W. A Thermalloy 6421B with a thermal resistance of 1° C/W would be a good selection in this application. It is important to use liberal coatings of heat conducting silicone grease on the transistor and associated insulating mica washers.

Summary

It is difficult to go into every detail required in order to produce a correct and exact power supply design to fit a driven set of requirements, so the approach that has been taken here is to include substantial safety factors so that the analysis and arithmetic could be simplified. It is recognized that lower rating components could be used, but then exact calculations and extensive analysis would be required. If the experimenter follows the design steps given here, the result will be a reliable, moderately priced power supply which most experimenters can easily build.

ESTIMATING POWER TRANSFORMER RATINGS

Junk box transformers sometimes have a way of losing their identity with time. It isn't too difficult to determine the voltages of the various windings, but current ratings are another matter. The best approach is to first get an idea of the overall power rating of the transformer. This is determined primarily by the amount of copper and iron in the transformer, so as an approximation the weight of the transformer will tell us roughly how many watts a transformer will handle.

With this idea in mind, several transformers of known ratings were weighed and their watts per pound determined by dividing the wattage rating by the weight. Here are some typical results:

Transformer Type	Watts	Wt/lb	Watts/lb
TV power	300	13.5	22.2
Old Radio power	132	7.5	17.6
Battery charger	600	24.0	25.0
Small radio power	40	4.25	9.4
Instrument power	20	1.6	12.5

Generally speaking, the larger the transformer, the more watts per pound you get. This is to be expected since a larger transformer is more efficient than a smaller one. The table may be used as a rough guide in determining the wattage rating of your transformer. Weigh the transformer and multiply the weight by the estimated watts per pound for that weight. Next estimate the current requirements for your application to see if the total number of watts is within the transformer ratings.

A word of caution when checking transformer windings on the AC line. *Always* use a test lamp in series with the winding until you are sure the winding connected to the AC

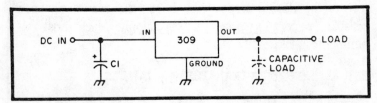

Fig. 1-17. Schematic of a typical three-terminal regulator circuit. If capacitor C1 should short, the output becomes more positive than the input, setting up reverse bias across the regulator's series-pass transistor, which can destory it.

line is indeed the primary. An unloaded transformer will show little or no primary current (lamp very dim) with the primary connected to the line.

SEVEN WAYS TO PROTECT AGAINST SMOKE

Monolithic IC regulators such as the LM309 et al are here to stay, as are the more sophisticated dual-tracking voltage regulators like Raytheon's 4194. They include a number of fail-safe features—most notably, short-proof circuitry and thermal overload shutdown, but in many cases, that's not enough protection. Presented here are some other possible failure modes, and how to avoid them.

Reverse Bias Across the Regulator

This accounts for a number of otherwise unexplainable regulator failures. To see why this problem occurs, examine the typical three-terminal regulator supply in Fig. 1-17. If the input capacitor should go rapidly to ground (through a short, for example) the output becomes more positive than the input, setting up a reverse bias across the regulator's series-pass transistor, which can destroy it. Adding a diode in series with the DC input, as shown in Fig. 1-18, can eliminate the prob-

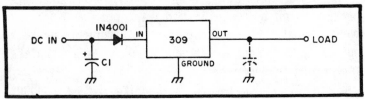

Fig. 1-18. Schematic of one method of protecting against reverse bias on the series-pass transistor, but this method adds series resistance to the circuit—a voltage drop.

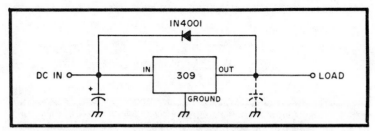

Fig. 1-19. Schematic of a typical three-terminal regulator with reverse bias protection. In this method the diode does not present any voltage drop and only conducts when the output is more positive than the input.

lem; but the diode does add a series resistance and consequent voltage drop. Figure 1-19 normally biased off because V_{IN} is greater than V_{OUT}; but should V_{OUT} become more positive, the diode conducts, dumping the current back to the input without going through the regulator itself.

Improper Polarity Transients at the Output

If a large negative transient hits the output of a positive regulator, or if a positive transient hits the output of a negative regulator, all kinds of troubles can occur. Unfortunately, transients riding the power supply line can be a fairly common occurence; once again a diode solves the problem.

Figure 1-20 shows a simplified diagram of a typical dual-tracking regulator. By connecting two diodes as shown, opposite polarity transients can do no damage. Any positive transients on the negative line shunt to ground through diode D1; D2 performs a similar function for negative transients.

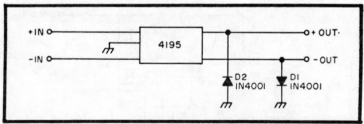

Fig. 1-20. Schematic of method of protecting the voltage regulator from line transients. This regulator, a dual-tracking type, is protected from line transients by inserting reverse-biased diodes in each supply line. When a positive transients is present on the negative line, D₁ will conduct, thereby protecting the regulator. Diode D2 performs a similar function on the positive line.

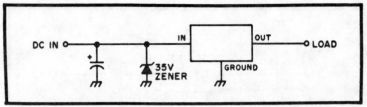

Fig. 1-21. Schematic of method of protecting a regulator from excessive input voltage. When the DC input goes above 35V, the zener conducts, protecting the regulator.

Excessive Input Voltage to the Regulator

The popular LM309 and several similar regulators are rated at a maximum 35V input, and that's for real. Anything over 35V can easily zap the regulator. Even if you're running around, say, 33V, a good voltage spike or upward change in the line voltage can cause the 35V figure to be exceeded. The best way to deal with this is to use a 35V zener across the input of the regulator, catching any possible overvoltage problems (See Fig. 1-21).

Excessive Voltage at the Regulator Output

Excessive voltage at the regulator output can happen in a couple of ways. The first is human error: test probes slip, a loose lead will brush up against an output terminal. Additionally, in some regulator systems external output transistors are used in conjunction with a low-power regulator (like the 723) to give higher output currents. When these external devices

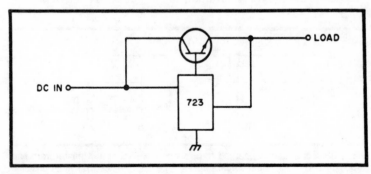

Fig. 1-22. Schematic of a typical regulator with an external series-pass transistor. If the transistor shorts, full voltage at the input of the regulator is present at the output.

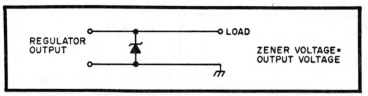

Fig. 1-23. Zener protection for the circuit in Fig. 1-22. By placing a zener across the output, excessive voltage will be shunted to ground.

fail (and they sometimes do, unless you're using one of National's LM395 blowoutproof transistors), the chances are excellent that they will fail as a short, rather than open, circuit. So what happens? Look at Fig. 1-22. The *full* voltage at the input of the regulator is now present at the output...not a healthy set of circumstances. The simplest way to deal with this problem is a zener diode across the output of the regulator, as in Fig. 1-23. Any excess voltage will shunt to ground.

A somewhat more thorough method of resolving this problem is with an SCR-resistor-zener diode crowbar circuit, applied to a simplified 15V supply in Fig. 1-24. As long as the output voltage remains at 15V, the SCR does not have sufficient gate current to trigger, and represents an off or high resistance state. But if the voltage on the output of the supply goes higher than 15V, the voltage differential between the gate and anode of the SCR suffices to turn the SCR on, causing a virtual dead short to ground and shunting any high voltage

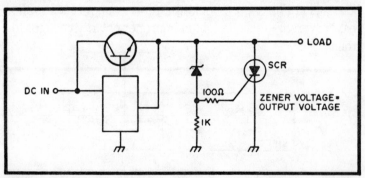

Fig. 1-24. Schematic of a typical regulator with a SCR-zener-resistor crowbar circuit attached for protection. When the output voltage is excessive, the voltage across the tener will be enough to trigger the SCR, producing a dead short. This keeps any high voltage away from the circuitry.

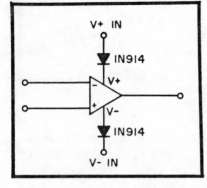

Fig. 1-25. Method of protecting op amps from power supply reversal. Placing two diodes in the supply lines insures that the op amp will not receive the wrong polarity.

safely away from the circuitry. Choose an appropriate zener for different power supplies; it should be the same as the desired output voltage.

Protecting Circuits from Power Supply Reversal

Although not a modification made to the power supply, protecting circuits from power supply reversal still falls under the heading of protection. As you may have already found out on the bench, reversing the voltage going to an IC can instantly destroy it. Simply placing two diodes in series with the op amp supply leads shown in Fig. 1-25 guarantees that even if you reverse the supply lines, the op amp will be safe from harm.

If you have a lot of ICs on a board, however, adding two diodes for each one can be a bit of a chore. In this case, try two diodes as in Fig. 1-26. Should the power supply lines to the board reverse, the diodes shunt improper polarity voltages to ground.

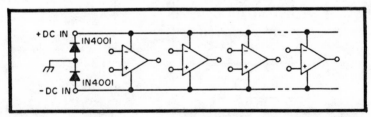

Fig. 1-26. Another method for protecting against power supply reversal. If there are many op amps to be protected, it can be costly to place diodes at each one, so just place your protection diodes on the supply lines with the op amps paralleled across them.

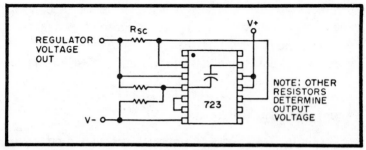

Fig. 1-27. Schematic of a typical 723 regulator with current-limiting protection.

Protecting the 723 From Excess Current

The 723, one of the most common regulator ICs, has a built-in current limit feature. Looking at Fig. 1-27, you'll notice that R_{sc} is the current-limiting resistor. You can derive its value from the formula $R_{sc} \cong 0.7/$max allowable current (for example, to limit current to 50 mA, $R_{sc} = 0.7/0.05$ or 14Ω). This feature not only protects the regulator, but the circuit under power.

Protecting Circuits from Excess Current

It is possible to limit current with a power transistor-zener diode arrangement for power supplies that don't have current limiting; an approach that involves a smaller parts count uses the current-limiting properties of the LM309. When hooked up as in Fig. 1-28, resistor R sets the current limit point. A good 309 will limit down to about 10 mA. Remember, though, that you are dissipating a certain amount of power through R, particularly at high-current values ($P = I^2 R$), so choose an appropriate wattage. The circuits shown limits from 10 to about 70 mA, a good range for experimental breadboarding.

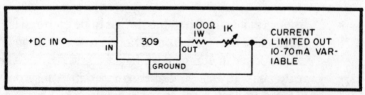

Fig. 1-28. Method of incorporating current-limiting for the 309 and others.

There you have it, seven ways to help protect your circuits and power supplies. If you use these various protective techniques, you'll find that the number of mysterious failures will go down, and that you will feel more secure about using the equipment you've made.

TRANSFORMERLESS SUPPLIES

In the past 15 years, electronics has shifted emphasis from electron tube to solid-state technology. This shift has opened new areas for innovation. Accordingly, the new generation of equipment performs more complex functions while using less components than previously possible. Old ideas can be reexamined in this light and may yield new results.

This section describes a capacitor-input, dual-voltage power supply that is suitable for low-power applications. It differs from the line-operated type found in the old AC/DC radios in that the hot side of the line is electrically isolated, the output is voltage regulated, and efficiency is high since the voltage is dropped across a reactive component.

Naturally the safety of such a device immediately comes to mind. Appliance shock hazard is measured in terms of a leakage current, which is that current that results when a wire is connected between ground and the appliance chassis. All appliances have *some* leakage current. In fact, any object that is even near house wiring has *some* voltage induced upon it; to demonstrate this, hook an ungrounded wire to the high-impedance vertical amplifier of a sensitive oscilloscope.

To be acceptable from a safety viewpoint, any leakage current must be minimized to a level consistent with current engineering practices. With this in mind, the capacitor-input power supply can take two different forms, depending on the particular application:

1. The special case, when the device to be powered is electrically insulated from all other equipment and any exposed metal chassis (such as driving a DC relay coil). In this special case the output may be allowed to float 115V above ground.

2. The general case, where one output terminal of the supply is electrically connected to a chassis or to some other equipment that has a chassis. In this case the capacitor-input power supply must be used either with a grounded three-conductor AC power receptacle or with an isolation transformer.

A low-voltage, low-current power supply can find many applications around the workshop. Many transistorized gadgets such as preamplifiers, frequency converters, electronic keyers, and speech processors need only tens of milliamps to operate. In addition, some IC projects require both positive and negative voltage supplies.

Bench-type variable power supplies are fine for testing equipment, but it is a shame to tie them up powering miscellaneous gadgets. Aside from using a battery, the alternative is to build a separate power supply for each gadget.

Typically, a 6V or 12V filament transformer is used to build the conventional power supply. Most hobbyist have several filament transformers in their junkbox; but they are often "boat anchor" types and are therefore esthetically unsuitable. Rather than purchase a small 1A filament transformer, a power supply can be built as outlined in Fig. 1-29.

As shown in Fig. 1-29, the 115V source is converted into a constant-current generator. This AC constant current is then split and rectified into a positive and negative constant current. The DC elements are then led into a combination constant-current sink and voltage regulator. This is a device

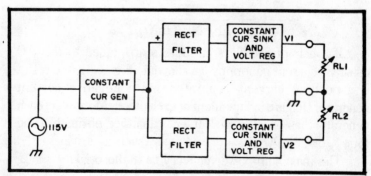

Fig. 1-29. Block diagram of transformerless power supply.

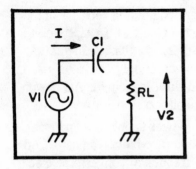

Fig. 1-30. An AC constant-current generator.

that will draw a constant current while maintaining a constant output voltage. The purpose of the constant-current sink is to limit the voltage at the output of the rectifier so that it will be independent of load resistors RL_1 and RL_2. This relaxes the voltage tolerance of the filter capacitor. The output voltages and currents are independent of each other.

Constant-Current Generator

When a capacitor and resistor are connected in series as in Fig. 1-30, they will approximate a constant-current generator as long as the reactance of the capacitor is much greater than the resistance. The current flow is determined mainly by the size of the capacitor and is independent of the resistor. This approximation is valid when V_1 is much greater than V_2. The current available is then:

$$I_{RMS} \cong V_1/X_C$$
$$I \cong 115 \ (2\pi 60C)$$
$$I \cong 4300 \ C$$

or

$$I/C \ (\mu F) \cong 40 \ mA/\mu F$$

If a pair of rectifiers and filters are added to this AC constant-current generator, as shown in Fig. 1-31, the result is a positive current generator and a negative current generator that are independent of each other. Moreover, each generator has an average current capability of one-half the total I, or 20 mA/μF.

The maximum power delivery to the load is

$$P = (I_1)^2 \ RL_1 + (I_2))^2 \ RL_2$$

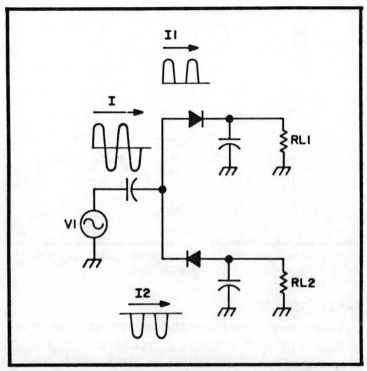

Fig. 1-31. Rectifier arrangement for independent positive and negative outputs.

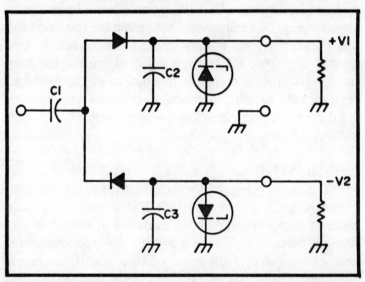

Fig. 1-32. Method of regulating the outputs from a low-current transformerless, capacitor-input, supply.

39

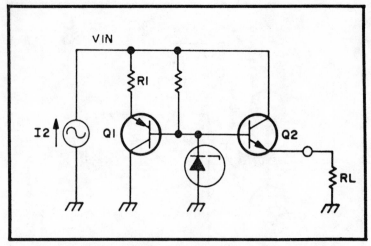

Fig. 1-33. For heavy current loads the capacitor-input power supply requires a transistor-type regulator rather than a simple zener.

Voltage Regulator

The voltage regulator must draw a constant current regardless of the load; otherwise the voltage at C_2 (Fig. 1-32) would rise up to 115V under light loading. For loads drawing less than 1W, a zener will probably suffice as in Fig. 1-32. But for loads drawing more than 1W, a transistorized regulator as shown in Fig. 1-33 will provide better voltage regulation and less ripple. When RL is very small, all the current I_2 goes through Q_2; when RL is very large, all the current goes through Q_1. If R_1 is chosen to be equal to the smallest load resistance RL to be encountered, then the voltage regulator will draw a constant current with no appreciable rise in input voltage.

Practical Examples

The circuit of Fig. 1-34 was constructed to verify performance. Note that a three-prong power plug and receptacle are required to assure a grounded output. The second ground prong is connected through a fuse to the grounded center conductor to guard against improper AC wiring. If the wiring is reversed, this fuse will disable the power supply and the neon bulb fault-indicator will light.

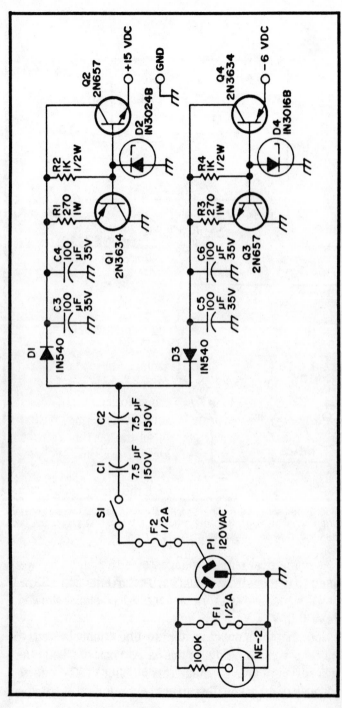

Fig. 1-34. Schematic of a typical capacitor-input dual-tracking power supply with +15V DC and −6V DC outputs. This supply will handle about 74 mA per leg.

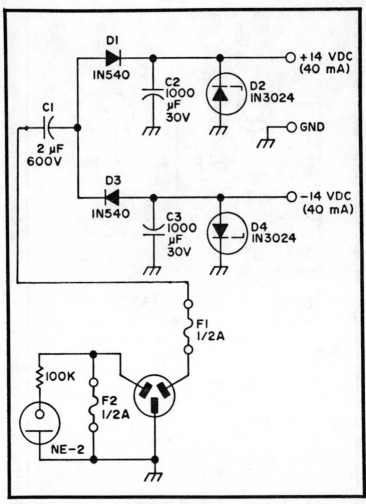

Fig. 1-35. Schematic of a low-current capacitor-input dual-tracking supply. Here the current capability is less than that of the supply in Fig. 1-34, only 40 mA, but it will work fine in certain applications.

Asymmetrical output voltages of +15V and −6V were selected to demonstrate flexibility. At currents up to 55 mA, the −6V signal had 0.1V ripple; the +15V signal showed a ripple of 0.05V.

Note that the input capacitor must be a nonpolarized type of sufficient voltage rating. In this case, a pair of CL33s were used in series to give a voltage rating of 300V. The effective capacitance of 3.7 μF in the current generator led us to expect

42

an average output current capability of about 74 mA (3.7 μF × 20 mA/μF).

The circuit of Fig. 1-35 is an example of a simple low-current supply. Again, more output current could be obtained by better filtering.

If a dual supply is not needed, one output polarity can be obtained by simply shorting one of the rectifiers to ground as in Fig. 1-36.

Two Conductor Wiring

Capacitor-input power supplies must be used with an isolation transformer if three-conductor AC power is not available. This becomes practical if different voltages are needed, since one isolation transformer can power several low-current power supplies. An isolation transformer can be built by the old trick of hooking two surplus filament transformers together as in Fig. 1-37.

Surplus DC relays are very inexpensive and are readily available. Their major drawback is that they usually require a

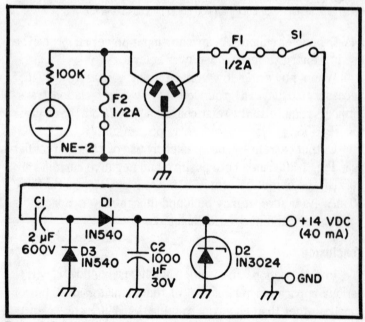

Fig. 1-36. Schematic of a single-voltage capacitor-input power supply.

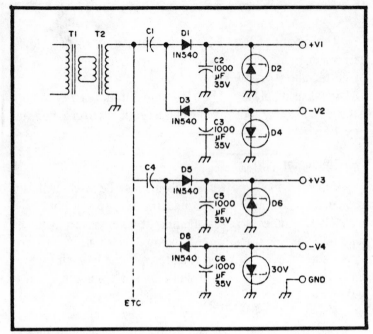

Fig. 1-37. Method of using back-to-back filament transformers when there is a three-conductor AC available. The filament transformers form an isolation transformer at a much lower cost than you could purchase one for.

24V DC power supply. A capacitor-input power supply can be used to energize most low-power relays.

When powering low-voltage relays, it is usually not necessary to have a grounded output. No isolation transformer is required if the relay coil is insulated from the chassis. Figure 1-38 shows a typical relay power supply.

Output current can be doubled by use of a bridge rectifier as in Fig. 1-39. Since both positive and negative currents are utilized, the output current is 40 mA/μF of C1, or in this case 80 mA. As before, it may be found that the filter capacitor is unnecessary and may be eliminated.

Conclusion

In this section, an attempt has been made to illustrate various application techniques of the capacitor-input power supply, rather than give construction details for any one project. Although the circuits described are certainly reproduce-

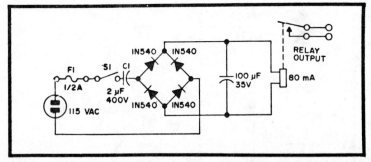

Fig. 1-38. Schematic of capacitor-input supply for operating low-voltage relays.

able, they are presented in their simplest form and could be improved upon by better voltage regulator and filter design.

Line-operated power supplies require greater care in wiring and an understanding and respect of the potential dangers involved. If you're unsure, stick with the isolation transformer.

USING TRIACS

It's easy to switch DC off and on with a transistor, but what do you use to switch AC? A relay would work, but this is bulky, and consumes power to perform its work. An SCR would work, except that it can only conduct for about half the applied sine wave, resulting in half voltage being applied to the load. Or you could use two SCRs, connected back to back. However, this configuration is already available in one package known as a triac.

A triac is much like an SCR, except that it will conduct for very near 360° of the sine wave. Triacs have three terminals, normally called main terminal one, main terminal two, and

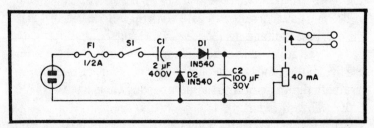

Fig. 1-39. Using a capacitor-fed bridge doubles the current-handling capabilities of the power supply.

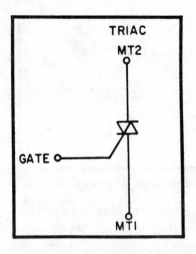

TRIAC
MT2

GATE

MT1

Fig. 1-40. Schematic symbol for a triac.

gate. MT1 and MT2 can be considered the normally open contacts of a relay, and the gate the coil of that relay (Figs. 1-40 and 1-41).

To activate the triac, trigger current is applied to the gate. This current can be either polarity DC or AC. Once the triac has turned on, current on the gate can be reduced to a much smaller value, called holding current. MT1 and MT2 will remain closed as long as gate current is at least this value. If gate current is removed, MT1 and MT2 are again isolated.

To operate a triac as a switch, we want it to be on for 360° of the applied sine wave AC. For this condition, gate current must be at least the value required to operate in *mode III positive*. For instance, the RCA type 40668 is rated at 8A, 200V RMS, and requires 80 mA gate current for this mode.

Let's suppose we have built a high-power linear amplifier, complete with safety interlock switches. These could be small microswitch types, wired as in Fig. 1-42. Resistor R is found by dividing gate current into gate voltage. Before the switch is closed, gate voltage will be 120V. Using 80 mA for gate current, we get a value of 1500Ω for R. To be on the safe side, use five to ten times the required gate current, or one fifth to one tenth the value of R. In actual use, any value from 100Ω to 500Ω will work well.

If R is replaced with a photocell, the triac can be turned on with a beam of light. When the cell is dark, its resistance is

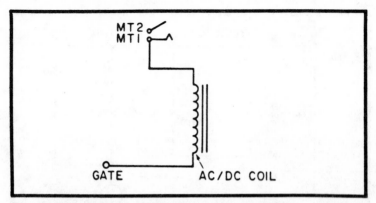

Fig. 1-41. Electromechanical equivalent of a triac is a relay. MT1 and MT2 act as contacts of a relay while the gate acts as the coil.

high, typically 1M to 3M. When in bright light, the cell resistance drops to somewhere around 150Ω. A small lamp, taped tightly to the cell, will allow full isolation of the signal voltage from the triac and AC being switched. Any CdS cell will do, and suitable small lamps, called grain-of-wheat lamps, can be found at most hobby shops. See Fig. 1-43.

The lamp could be driven by a small signal transistor in your terminal unit, and the triac could directly switch AC to the TTY motor, for autostart. A switch could parallel the photocell to allow running the motor as in a local loop.

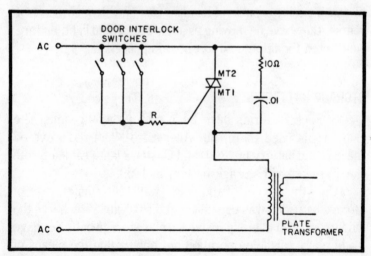

Fig. 1-42. Schematic of an interlock application of the triac.

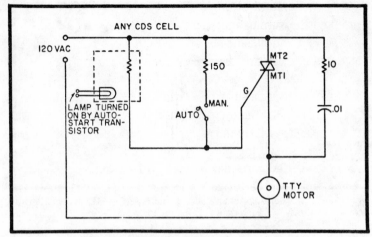

Fig. 1-43. Schematic of typewriter application of the triac. The triac is used to autostart the motor when a light is directed towards the photocell.

It is generally a good idea to put a resistor and capacitor in series across the main terminals of the triac to suppress any transients caused when the triac fires. Those transients could cause the triac to switch on or off randomly, or give sporadic operation. A 10Ω resistor and a $0.01\ \mu F$ capacitor will do quite well, although these values aren't critical. These are only a couple applications for triacs. They can be used anywhere a relay could be used, and are priced low enough to be competitive with a relay capable of handling an equivalent current. Of course, they have no moving parts, are small and light, and are well suited for printed circuit board applications.

STORAGE BATTERIES

Storage batteries have been with us a long time. The Babylonians used them, but Allesandro Volta rediscovered the galvanic battery in 1800, and Gaston Plante came up with the first lead-acid storage battery in 1859.

Until the advent of the solid-state age, amateur use of storage batteries was usually limited to trying to recharge the family car battery. Inexpensive surplus storage batteries have made really portable operation not only within the means of the average person, but a real pleasure.

Two types of storage batteries are generally encountered on the surplus market: lead-acid types similar to your car battery, and alkaline storage batteries. The most common alkaline batteries are the nickel-cadmium and the nickel-iron (Edison) cells.

When buying a surplus battery carefully inspect the case for cracks. Check the terminal posts for looseness or signs of internal damage. Remember, there may be a good reason for Uncle Sam to sell the battery! Buy the most recent battery possible (if they are dated). The negative plates of dry-charged lead-acid batteries tend to oxidize rapidly on exposure to moist air, so look for the ones that still have the seals over the filler plugs.

To get the best performance and life out of storage batteries requires careful maintenance and charging techniques. Dirt, corrosion and mechanical damage all shorten the life of a battery.

Terminal corrosion is best removed with a stiff fiber brush. After the corrosion is removed, wash the battery with water containing a mild detergent, and rinse with plain water. After replacing the connectors, coat the connectors and terminals with a light coat of cup grease or petroleum jelly to prevent further corrosion.

Connectors should be the proper size to provide maximum contact area with the battery terminals. Remember, lead has about twelve times the resistivity of copper! Always loosen and spread the connector for easy fitting. Never drive connectors on with a hammer or mallet as this may cause internal battery damage if it doesn't crack the case.

It is necessary to use completely separate sets of hydrometers and syringes for lead-acid and alkaline batteries. The alkaline cells are readily damaged by impurities, especially acid. The Army even insists that completely separate sets of tools be used.

The electrolyte should be kept at the recommended level. If electrolyte is lost by spilling, replace it with electrolyte. If the loss is due to evaporation or electrolytic decomposition bring the level up with distilled water. Don't use tap water as it may contain mineral impurities harmful to the battery.

Lead-Acid Batteries

These are by far the most common surplus batteries. The small 6V Sniper Scope units, and the cased 4V Marine Corps walkie-talkie units are most desirable. Most of these batteries come dry charged. Dry-charged batteries are activated by addition of sulphuric acid electrolyte. This is usually available at local battery shops. This is usually available at local battery shops. The most common specific gravity is 1.280. The electrolyte should be at least 60° F before filling. Fill slowly until the electrolyte fills about half the battery. Allow the battery to stand for 15 minutes. Tap it gently several times during this period to release the trapped gas. Continue to fill with electrolyte to the operating level. The battery should then be placed on a constant voltage charge (2.5V/cell) until it is fully charged.

Battery life is largely determined by the charge-discharge cycle (Fig. 1-44). Charging at too high a rate will cause the battery to heat up rapidly and gas excessively. This can damage the separators and buckle the plates destroying the battery. A temperature corrected hydrometer is desirable for checking electrolyte specific gravity during operation and charging. An approximate state of charge can be found from the specific gravity. For most high discharge rate batteries, a specific gravity of 1.280 represents full charge, and 1.100 completely discharged.

For constant-voltage charging, the voltage should not exceed 2.75V per cell. For constant-current charging the current should be approximately 10% of the rated battery capacity.

Alkaline Batteries

Nickel-cadmium and nickel-iron (Edison) are the two alkaline cells commonly found on the surplus market. They have the save positive plate material and use the same electrolyte (potassium hydroxide).

Unfortunately, most of the Edison cells found in surplus stores have been badly damaged. Thanks to design changes in certain antiaircraft missile systems there is a good supply of

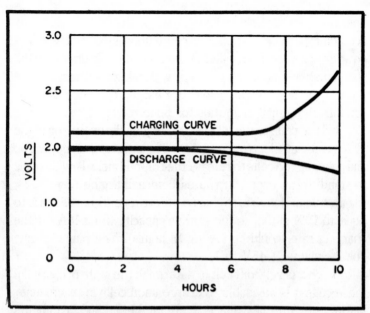

Fig. 1-44. Charge and discharge curves for lead-acid cells.

new, small nickel-cadmium cells. These cells are of two types, the hermetically sealed type containing electrolyte, and the dry-charged type. If you get the dry-charged type you will have to obtain vent plugs to replace the shipping plugs or you will rupture the cases on charging. Vent plugs are packed in the carton with the cells, but they have a habit of getting lost in surplus stores.

The potassium hydroxide electrolyte can be obtained from three sources: (1) surplus—the electrolyte is packed in cartons ready for use. Read the label carefully. It should have federal stock number FSN-6810-543-4041 printed on it (2) Chemical supply houses—it can be ordered as a powder or in tablet form. Order potassium hydroxide, reagent grade and specify that it is for batteries. You will need distilled water to dilute it to the proper specific gravity. *Warning*: always add the potassium hydroxide to the water, never the water to the potassium hydroxide! Proceed slowly. The chemical reaction liberates heat which can cause the solution to splatter if you aren't careful. Both the chemical and the solution are highly caustic. They will do a fine job of dissolving skin. It is quite poisonous so store it carefully out of the kids reach. Mix according to the directions, or approximately 9 ounces of the powder to a quart of distilled water. The specific gravity of the solution should be 1.32. Check it with a hydrometer, *but* not the one you use for lead-acid batteries! (3) Your local druggist can make up small quantities for you.

Fill the cells to the level mark, or just above the top of the plates using an electrolyte syringe of a large eye dropper. Tap and squeeze the cells to release the trapped air. Allow the cells to stand for at least 15 minutes before charging them.

A rule of thumb is to charge the cells for 15 hours at a rate equal to 10% of their rated service capacity (for a 10A cell the charging rate would be 1A for 15 hours). The final charging voltage will be 1.45V to 1.50V.

Since it is very difficult to determine the state of charge of these cells it is advisable to charge used cells in this manner before use. After charging the cells let them stand 2 to 4 hours before adding distilled water (if necessary) to adjust the electrolyte level.

The cells are rated at 1.25V under nominal load. Open circuit voltage of fully charged cells runs about 1.33V. The voltage when loaded to the 10 hour discharge rate ranges from 1.20V to 1.35V, depending on state of charge. The state of charge cannot be determined by measuring the specific grav-

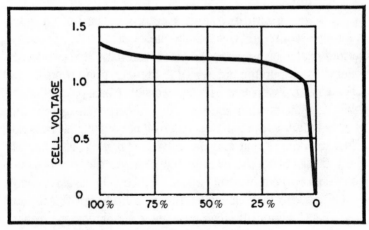

Fig. 1-45. Typical discharge curve for nicad cells.

ity of the electrolyte. One method is to measure the cell voltage under load. The standard 10 hour discharge rate is normally used. Figure 1-45 shows the approximate relation of cell charge to cell voltage. As you can see this is a very flat discharge curve. A more accurate way is to measure the current flowing into the cell when it is connected to a constant 1.50V source. If the current is less than 5% of the 10 hour rating you can assume that the cell is fully charged.

Paralleling of nickel-cadmium cells to increase the current capacity of the battery is not recommended. The internal resistance varies sufficiently so that it is difficult to prevent cells from overcharging, failing to charge, and from reversing polarity.

Charging

Two methods of charging are commonly used for either lead-acid or alkaline batteries. They are the constant-voltage and constant-current methods.

The constant-voltage method is the easiest to use, providing the charger has adequate capacity. Initial charging rates for a fully discharged battery are high, running from three to five times the 10 hour rate for nickel-cadmium batteries, and six times the finishing rate for lead-acid batteries. The advantage of constant-voltage method is that it is difficult to over charge the battery.

When using the constant-voltage method on lead-acid batteries you can take advantage of the quick charge technique used by service stations. A high percentage of the charge can be put back into the battery in a short period of time—if you are careful. The rule is to keep the charging rate (in amperes) less than the number of ampere-hours out of the battery. For example, if you have used the battery for 2 hours at a 5A rate you can charge it at a 10A rate.

When using the constant-current method care must be taken to prevent overcharging. For lead-acid cells this can be done by checking specific gravity of the electrolyte. For nickel-cadmium cells charge at the rated current until the cell voltage reaches 1.45V to 1.50V.

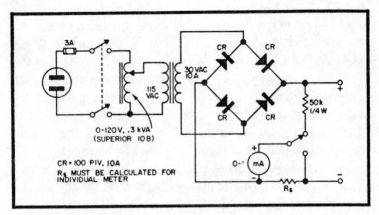

Fig. 1-46. Schematic for a 10A battery charger.

Figure 1-46 is the schematic diagram of the battery charged. It is a simple bridge rectifier device built into a case. A variable autotransformer provides continuous voltage adjustment, while the standard transformer provides isolation and voltage stepdown. All components except the rectifiers are mounted to the front panel. The rectifiers, on insulated heat sinks, are mounted to the cabinet. A meter serves a dual purpose, reading either voltage or current.

A three wire power cable is used to provide added safety. This unit was built for less than half the cost of a 10A commercially built home type charger.

FILAMENT TRANSFORMER CONVERSION

Are you having trouble findng a suitable filament transformer at what you may consider a reasonable figure for that new final RF amplifier you are building? Offered here is a method of exchanging your time and effort for such a transformer at a very modest cost. The problem can be resolved in three parts:

1. Determine the wattage requirements for the filament or filaments of the tubes to be used in the amplifier.
2. Find a transformer capable of supplying wattage.
3. Modify and rewind the transformer to supply the proper voltage and current.

As an example of the solution of part one, plan to construct an RF amplifier using four 811As. The filament in each

811A requires 4A at 6.3V. The filament wattage per tube will be 4.0 (amps) × 6.3 (volts) = 25.2W. Four tubes will require four times this wattage: 4 (tubes) × 25.2 (watts per tube) = 100.8W. Therefore any transformer used to supply the filaments should have a minimum rating of 100.8W for continuous service.

The second problem can be solved by visiting your favorite TV dealer or repairman to find out what junked TV sets with power transformers are available for bargaining. You should be able to pick up the transformer alone for a buck or less if you remove it from the old TV chassis. If the repairman does your TV servicing or the dealer has sold you some merchandise recently he may give you the complete set.

After you have acquired an old TV power transformer— or before, if you are given a choice of several sets—it is necessary to determine if the transformer has the wattage requirements to handle your tubes. As a first try, look up the manufacturer's specifications on the junker you have selected to determine if its power input requirements (wattage) are equal to or greater than the wattage requirements for the filaments of the tube you plan to use. If it meets this specification you may have found your transformer. If no information concerning the transformer you have selected can be found, it depends on how much of a gambler you are whether you wish to use it or not. It is advisable to select one with a known rating.

After a selection is made, check it over for shorts and that old tattle-tale burned smell. Carefully separate all the secondary leads and connect the primary to a fused 117V outlet. If it blows a fuse it may or may not be any good. However, as the primary is all that needs to be good, proceed with caution if it blew the fuse.

Assuming the primary has tested good, we can proceed to the third part. The modification may also be broken down into parts:

1. Factors that must be known: the voltage and current required from the secondary.
2. The numbers of turns of wire to supply the voltage required.

3. The size of the wire required to carry the filament current.
4. The actual rewinding.

The voltage and current requirements for the tube or tubes can be found in the spec sheet on the tube. The number of turns of wire to be used on the secondary can easily be found by using the following advanced mathematical gyrations. Count the number of turns on the transformer that are used on the 6V or 5V winding and divide the number of turns by their respective voltages. The resultant is the turns per volt for that particular transformer. The manufacturer has probably done a good job in obtaining this figure, so take his word for it. All that is necessary to do now is to multiply your desired voltage by the turns per volt ratio, and presto—the total number of turns for the secondary are found.

Now for the bug-a-boo that seems to rip everyone up—what size wire to use. The current carrying capacity of a wire is directly proportional to its cross-sectional area; however, conditions under which the wire functions will add a multiplier to this statement so that no hard fast rule can be made that guarantees that a wire size will carry so many amperes. The reason for this is predicated upon the wire's ability to dissipate generated heat under its operating conditions. For example, you would expect a wire in open air (such as a utility line) to dissipate its heat better than the same size wire (with length the same) coiled and encased in a transformer. Therefore, a larger wire will have to be used in a transformer than one used in open air conditions to carry the same voltage and current.

The size wire to be used is determined by the wattage and type of operations. By referring to Fig. 1-47, a value K can be determined for the particular wattage that the transformer is to handle. This value K multiplied by the filament voltage is equal to the necessary circular mils value of the wire to be used. For example—let's use the four 811As again—they required 6.3V and 100.8W. Looking on the chart, in Fig. 1-47 we find 100W; go up to the curve and across to a K-value of about 820. Multiply $820 \times 6.3 = 5166.0$ circular mils. From the wire tables we find that AWG #14 solid copper wire has a

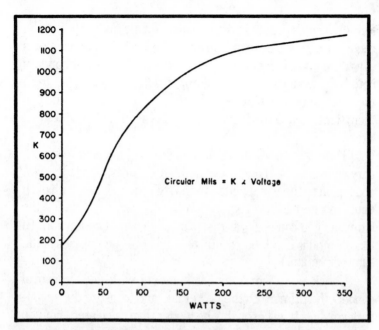

Fig. 1-47. Graph to determine the K-value of transformer windings. Multiply the K-value by the voltage of the secondary winding to find the circular mil cross section of wire.

circular mil cross section of 4107.0 and AWG #13 has 5178.0 AWG #13 is almost on the nose; however, it has been found that most motor rewind shops only stock even sizes of wire, so it may be necessary to go to the next even size larger or in this case AWG #12. AWG #12 has a cross-sectional area of 6530.0 circular mils and would probably operate slightly cooler. (It must be remembered that the larger the wire, the smaller the number of the wire.) A good grade of enameled copper wire, such as used in motor rewinding, is recommended.

Now that all of the paperwork is over we can get down to the rewinding job. Many manufacturers place the secondary winding on last, or on the outside, making it a simple matter to cut through the secondary layers with a hacksaw and remove them in a short time. Be careful and avoid cutting into the primary unless you plan to rewind it too. After the secondary is removed, place about three layers of plastic electrical tape over the primary windings if you don't have access to transformer varnish paper. It is now only a simple matter to fish the required number of turns through the transformer windows (this procedure is recommended if only the filament secondary is to be rewound). Be careful not to scrape the enamel from the wire on the sharp edge of the transformer's core—and pull them tight. Leave the leads long—you can always cut them off later.

There are two schools of thought on obtaining the secondary center tap: (1) count the number of turns, divide by two, and tap on to the wire, and (2) wind two secondaries with each equal to one-half the required voltage. Preference is for the latter, as you can easily be assured of a correct center tap. (If this method is used, be careful to wind both secondaries in the same direction.)

Check your work, and if it meets with your approval connect the primary to the power line and measure the secondary voltage. It will probably be higher than calculated, but you can expect it to drop when the tubes place a load on it. If the voltage is within 10% to 15% of the calculated value, connect the tubes to the transformer and check the voltage under load. If the voltage is too low, add a half turn or so until the voltage is correct. The reverse would be true if the voltage is too high. This should not be necessary if the above procedure was properly followed.

Place enough plastic tape over the winding to hold it firmly in place, and the job is finished. If there is a case for the transformer, slip a good grade of spaghetti over each of the protruding wires.

You'll find that this poor-boy transformer will not take a backseat to any of the commercial grades and you can't buy the experience you will have gained in rewinding it! Good luck!

ZENER DIODES

One of the many solid-state devices now available to the hobbyist is the zener diode. Properly used, it serves as a reference voltage source capable of delivering considerable current. Unlike a battery, its life is indefinitely long, although it must be supplied with a continuous current for many of its applications. This section contains the basic information required to intelligently design and use zener diodes.

For Instance

The number one application of zener diodes is probably DC voltage regulation for transistor circuits. Suppose you have just purchased a new 1N2974 zener. The catalog says 10W, 10V at 20% tolerance. Since you lack experience with zeners, you breadboard the intended circuit to pick up a few details on what zener regulators do. Perhaps the circuit is that of Fig. 1-48. Careful! The illustrated circuit is more suited to illustrative purposes than yours! You turn it on and this might be what happens:

A voltage measurement reveals regulation at 8.5V rather than the indicated 10V. You notice the voltage seems to be creeping up as you watch the meter. A signal tracer probe on point A finds lots of hum; but there is some at point B too. Perhaps if some current is drawn from point B the hum will decrease. A load test shows little effect on the hum except that under very heavy load the hum increases...some other noise too! Thinking about that, you smell smoke; it's the series resistor overheating. As you reach over to turn off the power supply, you notice the zener voltage seems to have

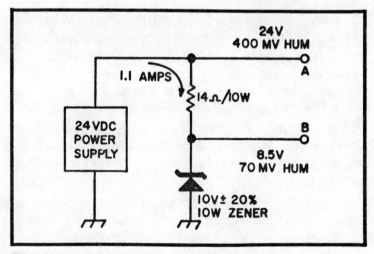

Fig. 1-48. A possible hit-or-miss zener regulator circuit.

dropped to zero. This reminds you that the zener has been operating without a heat sink. It's failed completely! Terrible. You can avoid the damage to wallet and peace of mind by using the following material. After reading, get out a pencil and some old envelopes or something, and design zener circuits. You'll soon catch the idea!

Zener Facts

Zener diodes are supplied in a large variety of packages. The controlling factor is how much heat must be dissipated. A majority of the zeners available are supplied in a diode-like glass package for up to 250 mW, a wire-mounted cylinder resembling a resistor or a silicon rectifier for the 1W size, a stud-mounting package for the 10W size, and a larger package resembling a power transistor for the 50W size.

A catalog search brought out hundreds of zeners rated from 250 mW to 50W. Some engineering books mention zeners as small as 50 mW (Cutler) to as large as 100W (Littauer). Operating voltages ranged from 3V to 200V or so. The least expensive were priced at 75 cents; $20 seemed to be the the upper limit, with a large variety in the $4 to $7 range. Price tends to increase with higher power rating and closer tolerance, but some good zeners are available in epoxy packages at low prices.

Some ordinary silicon diodes and transistors may be used as zeners. General Electric says some of their epoxy cased transistors can be used in this way. But most zeners are slightly special silicon diodes, designed to dissipate the fairly large amounts of heat produced in normal zener operation. Small zeners dissipate the heat along their leads or into the air; 10W and larger zeners are built like power transistors. The semiconductor material is brazed to a copper stud or surface which provides the route for dissipating excess heat into the required heat sink.

Only a small part of the package called a zener diode is actually the working element. This key piece is a semiconductor PN junction formed on a silicon wafer by a process involving some heat and considerable accuracy. In normal zener operation, the PN junction is biased opposite to the direction of easy conduction, at a voltage great enough so it conducts anyway. Sounds rough, but works fine. The arrangement is called reverse bias, and the zener always appears in the schematic with its arrow pointing toward the positive supply line. Figure 1-49 is a graph of current plotted against voltage for normal zener operation.

A small voltage invokes very little current. As the voltages approaches 10V, the current increases quite drastically toward some terrific value as the zener begins to act like a short circuit. If there is no current-limiting resistance in the circuit, the zener will promptly perish. This very rapid current upon voltage dependence gives the zener its useful voltage regulating property. The region of the curve in which the current first begins to rise is called the zener knee, and the normal operating region is called the zener plateau. The useful plateau is limited at one extreme by the current required to

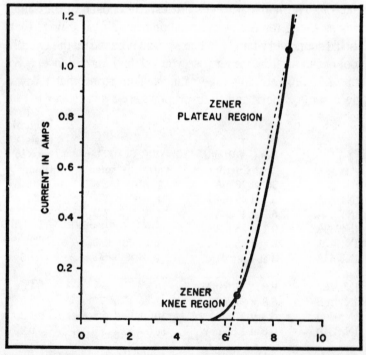

Fig. 1-49. Voltage-current curve of a 10W zener. The rise beyond the knee is so sharp the zener must be supplied from a current-limiting resistor or circuit.

keep the zener action alive, and at the other by temperature increase sufficient to destroy the zener.

The zener regulating voltage depends on how thick the PN junction is. If the PN junction is very narrow, the zener will regulate at a low voltage; we leave these details to the manufacturer. But depending on the structure of the zener it may show an increase or a decrease of voltage as it gets warmer! Zeners under about 5V will show a decrease in voltage, over about 5V a rise, with increasing temperature. A happy choice of voltage and current will give a zero drift: 40 mA at 4.8V, to 3 mA at 6V. Review the manufacturer's specs if real stability is required. Two or more zeners in series will show a smaller temperature voltage drift than a single equivalent higher-voltage zener.

Zener Specifications

All zeners are supplied with a voltage rating, and a tolerance. Like resistors, the standard tolerances are 20%, 10%, and 5%. The nominal values are usually chosen in the same way as those for resistors, resulting in voltage ratings that should sound very familiar. The system is based on the twelfth root of ten for 10% zeners, so you will find for instance, 3.3, 3.9, 4.7, 5.6, etc., voltage ratings. The listing that follows gives some zeners and their properties.

Type	Nom. Voltage Test Current	Watts	Tolerance	Dynamic Resistance in Ohms
1N4728A	3.3 at 76 mA	1	5%	10
1N4733A	5.1 at 49 mA	1	5%	7
1N4735A	6.2 at 41 mA	1	5%	2
1N4739A	9.1 at 28 mA	1	5%	5
1N4747A	20 at 12.5 mA	1	5%	22
1N4752A	33 at 7.5 mA	1	5%	4.5
1N957B	6.8 at 18.5 mA	0.5	5%	4.5
1N3016B	6.8 at 37 mA	1	5%	3.5
1N2970B	6.8 at 370 mA	10	5%	1.2
1N2804B	6.8 at 1.85 mA	50	5%	0.2
Z4XL6.2	6.2 at 20 mA	1	20%	9
Z4XL6.2B	6.2 at 20 mA	1	10%	9

If you are trying to regulate to a critical voltage, a germanium or a silicon diode may be placed in series with the zener (small increment) or in series with the load (small decrement).

The temperature drift problem can be minimized by keeping the zener cool. This conflicts with power handling ability but tends to guarantee long life despite experimental accidents. Like all semiconductor materials, if a zener PN junction gets too hot, the doping atoms begin to jump into new sites. This is very bad for the zener! Since the junction may withstand temperatures as high as 200° C, zeners are not remarkably fragile. But they cannot withstand the kind of overload even small power supplies can produce.

Zener wattage, as in any resistor, equals voltage across the zener times current through the zener. Check manufacturer's specs if much power is to be handled or if operating near maximum ratings. For breadboard and quick-and-dirty construction the fingertip test will do: too hot for a five-second fingertip touch equals too hot.

The zener's ability to stabilize and filter a power supply output is indicated by its dynamic resistance. Low dynamic

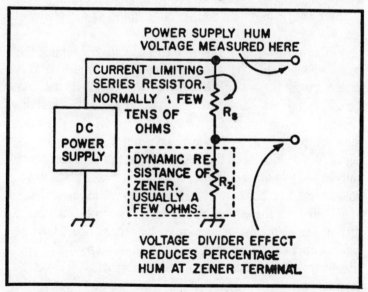

Fig. 1-50. Equivalent circuit for estimating how much a zener regulator will reduce the percentage and amplitude of power supply hum.

resistance is desirable. Suppose you wish to have power supply output stay within 0.1V of nominal in spite of 100 mA variations in current. By Ohm's law that works out to 1Ω: this is the dynamic resistance required. High wattage zeners, near 6V, have better dynamic resistance than any others; high-voltage zeners have very poor dynamic resistance but are not required for most semiconductor circuits.

The practical effects of dynamic resistance can be brought out by drawing equivalent circuits to show what's involved (see Fig. 1-50). An equivalent circuit is something used by engineers to simplify circuit problems. The dotted box suggests "we imagine the actual device acts like what's in there." From a hum viewpoint, the zener improves the situa-tion as if the entire circuit were a voltage divider. The upper resistor is the required series resistor R_s and the lower resis-tor is the zener's dynamic resistance, R_z. This is usually listed in the catalog entry. If the dynamic resistance is 1Ω, and the series resistor is 15Ω, the usual way of working out voltage-divider circuits tells us the hum will be reduced by a factor of 16. That doesn't remove it! Typical hum from a capacitor filter low-voltage supply is 0.5V to 3.0 to three volts. This would result in anticipated hum figures 30 to 200 mV; still plenty of hum.

Figure 1-51 shows the way to estimate how much the zener voltage will change under load. This figure is very different from the actual circuit so do not feel stupid if it's not clear! Try this with an actual zener when you've finished this section. Measure the zener voltage at a current near the knee. Measure it again at near maximum zener current. It will be higher. Now write down that higher voltage next to the "inside battery." If you draw current from this equivalent circuit, the voltage will drop because of losses across the inside resistor. This is just what the real zener circuit did: the small zener current corresponds to maximum load condition from the equivalent circuit. The voltage change divided by the current change gives the value of the series resistor. It will be the dynamic resistance again. This shows that knowing the dynamic resistance is very useful in reckoning the effects on regulated voltage of changes in load current.

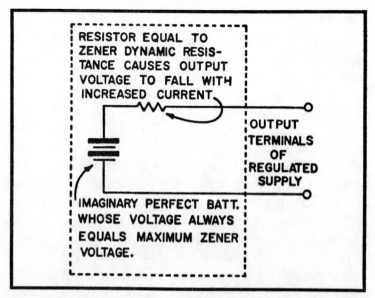

Fig. 1-51. Equivalent circuit for reckoning voltage drop with increased loading of a zener regulated supply.

Zener Noise

The useful ability of zeners to control circuit hum and noise is somewhat compromised by their natural ability to generate signals of their own. The zener regulating process is like an electrical discharge, and can produce similar noise. Good zeners produce very little noise, but some may become quite loud in the zener knee region. If your new circuit seems to be troubled by erratic frying and hissing noises, and this is traced to the zener regulator; the two solutions are increasing zener current to keep it out of the knee region, or the addition of a capacitor to take up the noise. Try 0.1 μF to start.

Varactors are reverse-biased silicon diodes. Since zeners are also reverse-biased silicon diodes, do they have an associated capacitance? They certainly do; it may be as large at 0.01 μF. This capacitance is unimportant in normal operation, and probably has a beneficial effect in reducing zener impedance at high frequencies. Perhaps this capacitance can be put to other uses! A possible application would be oscillator tuning; perhaps a cheap zener would work better than a cheap silicon diode.

Zener Regulator Design

All zener circuit designs depend on the same basic facts of the zener's voltage, tolerance, wattage rating, dynamic impedance, and perhaps temperature drift. The worst case from the zener's viewpoint is DC power regulation, with a steady current supply from a voltage source capable of destroying the zener if the series resistor fails or is shorted. Zener regulators are very useful and deserve a close examination.

Only professional engineers should design zener circuits to operate at the limits of the zener's capabilities. The hobbyist, by leaving generous margin for error, can simplify the design problem to a point where only the simplest math and less than complete information on the zener's capabilities will be sufficient for a reliable design. Apart from possibly very serious misunderstandings on the hobbyist part, the major troubles that might arise are the relatively generous tolerances on most inexpensive zeners, the wattage problem, and voltage change under load resulting from dynamic resistance.

The design process commences with finding, some working figures. Refer to Fig. 1-52, a design sketch made just before carrying out the following procedure. Power supply voltage, zener voltage, and load current information are collected, along with their estimated maximum and minimum values. A class B audio amplifier of a few watts capability could account for the fairly large current variations shown in the diagram; this is a rather extreme case. But it could happen! If the load were a receiver small-signal circuit, an oscillator, or a transmitter VFO; the load would be practically constant. After collecting this information from figures, estimates, educated guesses, and by breadboarding, the circuit is designed on average values. When a series resistor and a zener are chosen, their anticipated properties are checked against possible extreme conditions of voltage and amperage.

Variations in load current are made up by opposite variations in zener current. Minimum zener current flows when maximum load current is taken from the regulator. This adds up to the first requirement for the zener: it must carry a current greater than the anticipated load swing. The regula-

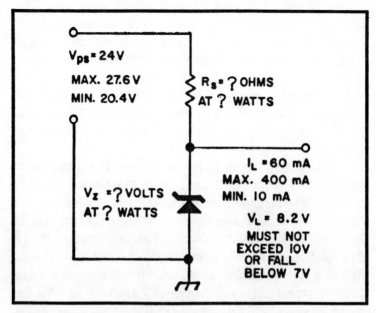

Fig. 1-52. Designing a reliable zener regulator. The question marks indicate values to be worked out.

tion fails if the zener is starved into the knee region, and if the zener is overheated by excessive current, catastrophe is likely. Often a well-placed large capacitor in the load circuit will absorb the drastic swings. The zener required in Fig. 1-52 must carry more than the swing of 390 mA. If the minimum and maximum loads were both greater by 1A, the swing would remain the same and so the zener minimum current requirements would be unchanged. However, with increasing loads flowing past the zener, a point may come at which turn-on or turn-off of the circuit results in transient overloading of the zener.

A safe minimum zener current is 10% of its maximum rating. A fresh zener accompanied by manufacturer's specs may be used at much lower levels, taking the specs as a reliable guide. For surplus zeners the 10% lower limit is recommended unless a test shows the particular zener will stand further starvation. Since the voltage requirements are already determined, the choice is made on the basis of wattage. Don't be afraid to use an oversize zener; it's safer and the larger zener will have a lower dynamic resistance. The absolute minimum wattage required equals the maximum possible voltage across the zener times the maximum expected current through it. There seem to be no 5W zeners; for most regulator purposes the choice is limited to 1W or 10W. Or 50W if your wallet can stand the drain. The zener in Fig. 1-52 should carry about 450 mA under minimum-load conditions; at 9V that works out to just under 4.5 watts. A 10W zener is indicated.

Now series resistor Rs can be chosen. The voltage at its upper end is fixed by the power supply, and at its lower end by the zener. The series current is the maximum load current plus the minimum zener current finally decided on. In the case of Fig. 1-52 this was a total of 450 mA. This same current flows if the load drops to 10 mA, since the zener now takes 440 mA. Knowing the voltage across the resistor, 24V minus 8.2V or 15.8V, it appears, that a resistance of 35Ω and about 7W is the minimum value.

This leaves no margin for error. A better choice is an adjustable wire-wound resistor, 50Ω, 25W. More than half the resistor will be in the circuit. If only half or 25Ω were used, it would still be rated at 12.5W, so that with this choice the success of the design seems probable.

Now we return to the zener. Any electrical slop in the design can be taken up by the resistor, and we know that a 10W zener is required. Referring to the catalog, we find a 1N2973B, 9.1V 5% zener. We expect the circuit can withstand the possibly slightly high voltage. If not, we'll change the circuit. We choose a zener at the high end of the range because its voltage will drop under load: a dynamic resistance of 2Ω means 2V drop per amp decrease in zener current, or in the case of Fig. 1-52 the voltage will swing over a range of 0.23V. Adding to this the 0.45V possible error due to zener tolerance, and adding that to the 9.1V nominal zener rating gives about 9.8V as the largest we should expect to see in the circuit. Subtracting the same figure from the 9.1V nominal figure gives a minimum of 8.4V in the case of an extremely low valued zener. These results are within the previously decided requirements.

There is still the question of changes in power supply voltage. What happens if line voltage changes drastically? This cannot be answered simply; some line voltages are more changeable than others! In Fig. 1-52 a 10% variation either way was just picked out of the air. This is probably large. Now if the zener voltage is fixed at 10V, which it will never quite reach in the actual circuit, and if the supply voltage drops to 20.4V, the series resistor being at 35Ω with 10.4V across it now passes only 335 mA. Not enough, so we adjust the resistor to 23Ω, and it now passes 450 mA. But now we must try the other extreme: the power supply voltage rises to 27.6V and we suppose the zener to be an 8.4V type. Then we find 19.2V across a 23Ω resistor. That is about 840 mA. The zener won't overheat if it is properly mounted, but the resistor must dissipate 16W. Since less than half of it is carrying current, we must go to a smaller resistance at the same wattage in order to get enough dissipating surface. The choice is a 25Ω 25W resistor, still adjustable. Or if you're a little apprehensive about going that close to the limit, a 50Ω 50W resistor can be purchased at an 83 cent increase in price.

That completes the design. This process avoids the following kinds of grief: zener failure due to overheating, zener voltage out of specs, circuit loads zener regulator into the knee region, and regulator fails due to starvation or overheating at extreme line voltage values.

Correcting Zener Voltage

The combined effects of high prices and large tolerances are hard to beat! But a resourceful amateur need not lose a project just because his nearest zener isn't quite near enough. Zener voltage can be adjusted up or down by correct use of small germanium or silicon diodes. The price is a slight increase in dynamic resistance and in temperature drift. Some knowledge of the properties of forward-biased diodes is required.

A silicon or germanium junction diode, carrying a forward current of 10 to 100 mA, depending on its size, has properties very like those of a low-voltage zener. In fact, there are no

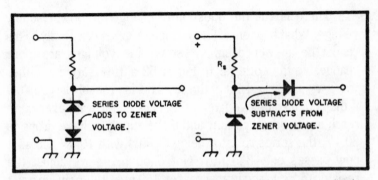

Fig. 1-53. The effective regulating voltage can be adjusted by correct application of ordinary germanium or silicon diodes. (A) By placing the added diode in series with the zener, the effective voltage increased by the forward voltage drop. (B) By placing the added diode in series with the load, the effective voltage is decreased by the forward voltage drop.

zeners under about 3V, and diodes are used in just this way to fill the remaining gap down to near zero. Beyond the early stages of conduction, a few microamps or mils, the diode voltage changes very little with current. Its dynamic resistance is quite low. Diodes can be used in zener circuits as if they are little batteries, to achieve a slight increase or decrease in apparent zener voltage. The voltage measured across the zener itself is not affected.

The voltage at which the diode regulates depends on its material: germanium or silicon. A germanium diode well into conduction will show a stable voltage of around 0.3V; a silicon diode regulates above 0.7V. A transistor base-emitter or base-collector junction could be used in place of the real diode; it's a PN junction too and will show the same behavior.

To achieve a small increase, the diode is placed in series with the zener, as shown in Fig. 1-53A. The reverse-biased zener and the forward-biased diode point in opposite directions. The drawing is arranged so that positive current—a convention—flows down. Figure 1-53B shows the diode in series with the load so that its voltage subtracts from the zener voltage. They seem to be pointed in the same direction, but the current flows against the zener and with the diode. This is certainly confusing and will require some careful thinking. Try it; make your mistakes on a breadboard where they show clearly and are inexpensively remedied.

Amplified Zeners

Being relatively high priced and having rather large tolerances, zeners may seem rather useless to many hobbyists. But a small, inexpensive zener can be combined with a transistor, making a simple two-terminal circuit that will stand in very well indeed for a 50W or even larger zener. This particular transaction shows an unusual measure of profit: besides greatly reduced price and substantial easing of power limitations, dynamic resistance may be improved and becomes little affected by using diodes in series with the zener to build up its voltage. The effects of temperature upon voltage are increased but this will rarely be important. The current handling ability is multiplied by the transistor's beta but the temperature drift is only that of the individual diodes in series.

For instance, a Texas Instruments 2N251A at $2.25 plus a General Electric Z4XL6.2 at 75 cents adds up to $3.00 for a shiny new, somewhat adjustable zener, rated about 50 watts depending on the beta of the transistor. This is comparable to the 1N2804B, priced at $10.65. That's what makes amplified zeners interesting!

The complete circuit is shown in Fig. 1-54. It does seem rather bare in comparison with most transistor circuits, but eveything that's really required is there: one zener and one transistor. This two-terminal circuit closely resembles the

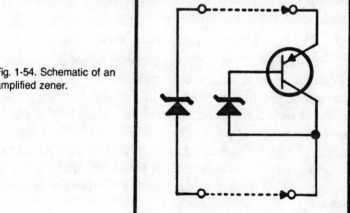

Fig. 1-54. Schematic of an amplified zener.

emitter follower regulator, and if a resistor were added from the zener diode/transistor base connection up to a higher voltage to ensure liberal zener current, it would be an emitter follower regulator. But the resistor can be omitted if the amplified zener's knee region is avoided, and then the current divides between the transistor and the zener according to the beta of the transistor.

One such amplified zener can be constructed from a ZN538A transistor and a 6V 1W zener. This one costs an estimated 50 cents and gives very good test results. Its knee region seems to end at about 4.2 mA and the actual zener diode is not overheating at 1.6A regulator current. A current increase to about 1.6A boosts the voltage from 5.6V to 6.8V, for an average dynamic resistance of 0.67Ω. Using the voltage-divider method and hum input-output measurements, its dynamic resistance at 1.6A is 0.59Ω. The very best zeners are little better than that. There is one hidden pitfall: the transistor's leakage current increases with temperature. This extends the knee region to higher current values.

This is how the current division works out. There are only two terminals; the current must go in one and out the other. It takes two routes in between. Suppose the base-collector zener is carrying m milliamperes. The transistor base-emitter junction supplies this current, and as a result an additional current, beta (β) times larger, flows from emitter to collector of the transistor. The total current I is the sum of these two, so that we write

$$I = m + \beta m = m \ (1 + \beta)$$

The one plus beta in the parentheses is not particularly different if we leave out the one, provided the beta is greater than ten or twenty. It usually is in a usable transistor; the difference between ten and eleven is 10%, small by electronic standards. For most purposes it's simpler yet true enough to say all of I goes through the transistor, the circuit regulates at the zener voltage plus the transistor base-emitter voltage, and the zener heating current is I divided by beta. The error is trivial.

For example, the Z4XL6.2 is rated at 1W, the 2N251A at 90W, and suppose a beta measurement under approximate operating conditions gives a result of 50, well within specs. Remember that beta is quite evanescent, depending upon collector current in addition to great variations between transistors of the same type! The maximum allowable zener current is 160 mA, since 0.160A times 6.2V equals the rated 1W. Then 50 times 160 mA gives a maximum of 8.4A current. The amplified zener regulates at 6.4V, since the live germanium transistor will show about 0.2V from base up to the emitter, which is added to the zener voltage. If you really want to dissipate 50W, the transistor should have higher beta or a 10W zener should be used; stay away from calculated limits!

The dynamic resistance of the amplified zener will be the inside real zener's dynamic resistance divided by the transistor beta. This works out to one-fiftieth of 9Ω, or 0.18Ω. This value is so low that the power transistor's characteristics may enter into the final result; the final value will still be well under an ohm. This result is not appreciably spoiled by adding series diodes to pad up the zener's apparent voltage.

As the current through the amplified zener is reduced toward zero, the real zener and the power transistor both weaken. The combined effects are rather uncertain, so that breadboarding with the actual components is a good, safe practice. Find the knee by measurement; remember that a capacitor across the zener will reduce its noise generating capabilities.

Other Zener Applications

Zeners do not go very well in parallel. One will tend to hog the current. There is no need for parallel zeners anyway; an amplified zener will do a better job. But zeners can be connected in series to provide two or more regulated voltages. And in this case the ground can be between the zeners, rather than at one end of the power supply. If you ground both points, it won't work!

The design of switching circuits is considerably simplified if the usual collector voltage supply is supplemented by a

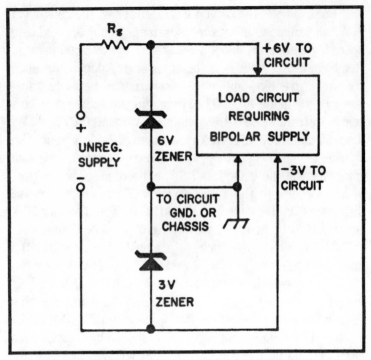

Fig. 1-55. A pair of zeners in series to provide both positive and negative voltages with respect to circuit ground, using a single power supply.

lesser voltage of opposite polarity. The second supply is used to drain off unwanted leakage currents, turn diodes and transistors hard off, and for other applications. This relatively slight increase in the designer's armament eases many tough circuit problems. By using a pair of zeners in series, the desirable pair of voltages can often be obtained without going to the time and expense of a separate second power supply and all its problems of cost, space, weight, and regulation. Figure 1-55 shows how simple this arrangement is.

Without a load circuit, the same current flows through both zeners. If some current is side tracked around either zener, the voltage across it remains constant. The regulation isn't disturbed at all if some current is taken out around both zeners, and this is the normal application. The usual considerations about starving and overfeeding zeners are applicable here, and the beginning designer should remember that the two supply circuits don't necessarily require the same currents at the same time.

The zener diode shows little promise as a limiter and no schematic for this application is included. Ordinary diodes are distinctly superior. Zeners require too high voltage, 3V or more. In normal circuits that's in the high-power range. The clipping should be carried out well before the signal gets this large. Also, the zener will clip at normal silicon diode levels as soon as its Pn junction is forward biased. Additional diodes would be required to prevent this, unless unsymmetrical clipping were intended. Zener clipper circuits appear impractical.

The only remaining field in which vacuum tubes retain some superiority to solid-state amplifiers is large-signal RF power amplification. The zener diode can fill a very useful spot here. It can replace the cathode bias resistor, offering a bias voltage quite independent of tube current. Figure 1-56 shows a zener in this application.

Because the zener acts like a battery, most of the high voltage is taken up by the vacuum tube. The zener merely guarantees the bias. It never runs down or emits corrosive chemicals, and has a lower internal resistance than the batteries used for this application in the old gear described just

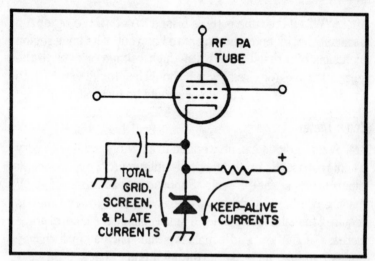

Fig. 1-56. Zener diode biasing for an RF power vacuum tube. Bypass capacitor recommended, appropriate for operating frequency.

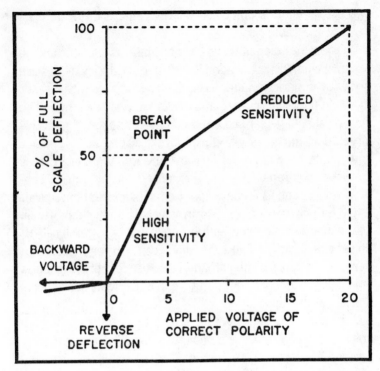

Fig. 1-57. How an improved meter might be calibrated to show small and large voltages with comparable accuracy. Reverse voltage does not bang the meter backwards.

after WW2. If the tube is to be biased to cutoff, the zener can be supplied with enough current to keep out of its knee region by means of a resistor up to the high voltage or over to the adjacent transistor circuit which should be providing the RF to the power amplifier. The zener should be bypassed for RF.

Zener Meter

A zener diode can be wedded to a meter circuit with very useful results. To understand the utility of this match look at the usual linear meter scale. Suppose it reads to 20V. A 1V reading will be way down at one end, and a lower range is required to make it readable. A small change huge percentage increase at the low end equals in scale space a small change tiny percentage increase at the high end. That's not a very equitable distribution. For example, it would be convenient to check transistor emitter-base and emitter-collector voltages without changing ranges.

The required benefit is achieved if the circuit can be made to show variable meter sensitivity. Figure 1-57 shows a realizable result, detailed below. The first half of the meter scale is taken up with the 0 to 5V range. The 5V to 20V range occupies the second half of the scale, without switching. The poor sensitivity to voltage applied in the wrong direction is a valuable by-product of scale tailoring with a zener diode.

At first glance this circuit (Fig. 1-58) appears to have been designed by a network expert. A closer look reveals that the values of the resistors may be deduced, one at a time, by thinking out the inside requirements of the circuit. R1 and R2 will fall first. If the meter is to read 5V at half-scale with 5V applied, R1 plus R2 must come to 200K since this will pass the required 25 μA. The Lafayette meter's resistance of 1K is

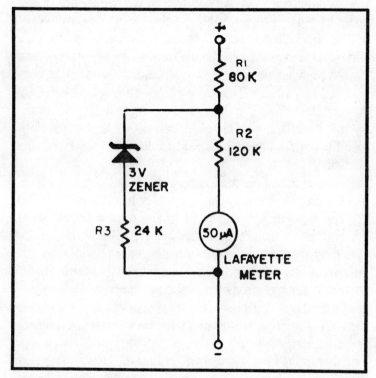

Fig. 1-58. A circuit that will produce the characteristics shown in Fig. 1-57

insignificant in comparison to this value. Supposing at 5V the zener hasn't quite broken down, it must have just 3V across it. The voltage across R1 must be 2V, and at 25 μA the resistance must be 80K. That leaves 120K for R2 since the pair must add up to 200K. We already know the zener; the problem is two-thirds solved.

Now we proceed confidently to the determination of R3. At 20V applied the meter reads full scale, therefore is carrying 50 μA. This current through R2 must bring the junction between R1, R2, and the zener to 6V. Now there must be 14V across R1 and that yields 175 μA through it. The meter gets 50 μA; 125 μA pass through the zener and its series resistor. We have it! From the junction through the zener we lost 3V, by design; the remaining 3V μA at 125 μA fixes R3 at 24K.

A breadboard check shows that the circuit behaves about as shown in the graph. This graph preceded the design, and the actual circuit is influenced by the characteristics of the zener in its knee region. Because the zener comes into conduction gradually as applied voltage is increased, rather than abruptly, the actual scale change from steep to flatter occurs along a rounded curve. A new calibration scale must be constructed empirically. That is, each point must be located by applying the indicated voltage and marking or listing the resulting meter deflection. This good idea needs further development; it requires enough current to disturb many transistor circuits.

The constant-current generator circuit closely resembles the amplified zener. Only a resistor has been added. But the constant-current generator guarantees a certain fixed current, rather than the amplified zener's reliable voltage. Its operation depends on the resistor; the zener provides a reference voltage and the transistor, acting as an emitter follower, holds that voltage across the resistor. The resulting current, determined by Ohm's law, is independent of voltages applied to the outside circuit terminals if the transistor is biased into its operating range.

A working circuit is shown in Fig. 1-59. Remember that the power dissipated by the transistor is determined by its

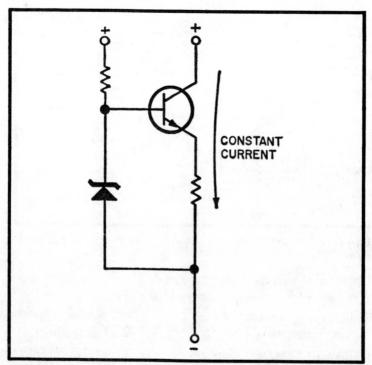

Fig. 1-59. A zener diode combined with a transistor to make a constant-current regulating circuit.

collector current and voltage, not by the values at the rest of the circuit. As in the amplified zener, if the transistor beta is large enough the zener current may be ignored. The computation proceeds in this way: the 6V zener fixes the voltage across the resistor at 5.8V, because 0.2 is lost across the base-emitter junction of the germanium transistor. If a silicon transistor were used, the resistor would see 5.3V. Since a current of 100 mA is to be guaranteed, the resistor must therefore be 58Ω. A fixed current of 10 mA would require a 580Ω resistor since the voltage across it is held constant. But that might not work so well since the zener could be starved for current; perhaps the zener could be biased elsewhere and its voltage carried over to the transistor base.

This is an excellent circuit for eliminating hum. The hum current cannot pass the constant-current circuit. But before this circuit can be put to work in a usable power supply, it must

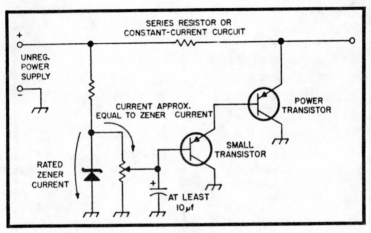

Fig. 1-60. Schematic of a very close relative of the amplified zener. It acts as a variable zener.

be provided with an appropriate load. The fixed current will generate a large voltage across a large resistor, a smaller voltage across a smaller resistor, and a zero voltage without blowing up anything across a short. This fail-safe feature can be retained while correcting the terrible regulation problem by adding a zener regulator. The constant current is just right for biasing zeners; it is inserted in place of the usual series resistor, and a really good supply results.

Finally, the last circuit is a realizable substitute for a continuously variable zener. It looks very much like a Darlington pair used as an emitter follower (see Fig. 1-60). A current of 10 mA or so from a zener regulator puts a fixed voltage at one end of a pot, decreasing to zero at the ground end. This voltage is stable if very little current is drawn. But the current required by the Darlington pair to regulate at a certain voltage will be the through or zener-like current divided by the beta of the first transistor, the result divided again by the beta of the second transistor. A milliamp will determine 1A to 10A. The illustrated circuit will regulate from about 1V to 15V. The capacitor is required to take out hum coming around through the zener reference voltage source. A supply using the two circuits above shows regulation as good as simple feedback-regulated supplies, combined hum and noise of about 0.6 mV, and adjustability over a wide range.

Surplus Zeners

Zener diodes are available at prices well under par from several sources. The routes by which these zeners enter the surplus are not at all apparent, but it seems that in many cases these zeners are rejects having no place in any electronics gear hobbyist or otherwise. It also appears that some suppliers do not test their zeners as well and carefully as advertising statements seem to indicate.

Assorted zeners from one supplier were tested for zener voltage and dynamic resistance. Most tended to regulate in the general 20% region, but a few were drastically off. Many of these zeners, 10W stud mounting types priced under a dollar each, had fairly high dynamic resistance. Perhaps that is why they were available. A second collection, about 30 assorted zeners adding up to the attractive price of $10 plus shipping, appeared considerably less economical after careful checking and tests. Some were mounted backwards in their cases, many showed poor regulation, a few were phenomenally noisy, others did not zener at all, and one had a broken lead. The more expensive varieties did not seem to average any better than the cheapest ones. There is a moral here. If you are going to use surplus zeners, check regulating voltage, dynamic resistance, and noise characteristics of each zener before you put it in that nice new circuit. Don't take it on faith; the chances that it is not as indicated may be as bad as one in two.

This experience suggests that the most effective way to buy is to purchase new stock zeners, or else test before buying. It may help to initiate a general practice of testing zeners promptly upon receipt, and returning bad ones to the supplier. Be certain the test is correct. Or perhaps you have found a good source of tested surplus zeners; if so, make the most of it and tell your friends. A zener is a zener, and it's the device, not the label, that is required in the circuit.

Testing Surplus Zeners

A batch of surplus zeners can be tested most effectively if the operation is performed in several steps. The first pass eliminates the obvious duds, the second sorts out the remaining zeners into broad voltage ranges. A third, perhaps, determines if a particular zener can be used in a specific application.

Several instruments are required for complete testing. Also a few resistors and clip leads, a place to work, some scratch paper, and marking paint are required. A high-sensitivity multimeter of a DC VTVM serves for voltage measurement. Another multimeter or a milliammeter provides for current measurement. An optional AC VTVM is useful for checking dynamic resistance by the hum voltage divider method. A signal tracer will serve very well for detecting the slight hiss a few zeners show in the knee region, or the raucous racket at higher current levels indicating the zener should be discarded. Finally, a magnifying glass assists in detecting mechanical faults on the surface of the package.

The surplus market is low man on the totem pole. It's quite safe to expect a specific zener has something wrong with it, which brought it to the supplier and then you. The testing operation is a sort of detective game played to find out if the fault will or will not interfere with its use in a piece of ham gear. This game can be played most productively if the goals are known. First, does it show semiconductor properties at all? Second, what do they seem to be? Finally, does a closer inspection show they are really there, and that obvious faults are absent?

Modern technology and the manufacturers have conspired to make this game more difficult than it might be. A given zener may be a double-anode device, usable in either direction. Or it may be a zener and a diode, practically the same thing but rather different in intent. And there is a chance it is an amplified zener: a zener and a transistor in one package. The amplified zeners seen so far have been high-power devices, but this may change any time.

A first inspection serves to eliminate broken zeners, ones with bad leads, cracked cases, and other faults. An obviously abused condition is certainly grounds for rejection. At this time the wattage can be estimated by comparison with known zeners and catalog descriptions. Low-wattage accurate zeners may be placed in large cases for better temperature control; high-wattage zeners are indicated by the provision of some means for mounting to a heat sink.

Then the power supply is set up with its negative output terminal to ground. A resistor is placed in series with its positive terminal, chosen to limit the current to near 10 mA. A 40V supply would require about 4000Ω, anything over a half-watt would do. The zener goes between the output end of the resistor and ground. Regardless of its condition it cannot receive more than 10 mA; this is a safe arrangement.

The first test is to measure the lowest voltage across the zener at this current, trying both directions. If there is a polarity mark or band, the least voltage should be seen when the band or cathode end is toward negative ground. If the voltage is under about 0.6V, the device is not a zener and further testing is not required. If it is in this range, and if doubling the current by halving the series resistance from the supply produces only a small increase in voltage, the device is showing proper characteristics for a forward-biased silicon PN junction. If this cannot be achieved, it may be a faulty device, or it may be one of the more complex varieties, mentioned but otherwise carefully avoided in this section.

A breakdown test is now appropriate. The cathode end is turned toward the positive supply, and a measurement of voltage gives the approximate zener regulating voltage. If the resulting voltage is the power supply voltage, then no current is flowing and the device may be a rectifier whose inverse voltage, or a zener whose breakdown voltage, exceeds that available. A current doubling should, again, have very little effect on the stabilized voltage. This test indicates that the device shows zener characteristics.

Knowing the approximate zener voltage and wattage, the supply circuit can now be revised to bias the zener to anticipated normal operating conditions. At this time the dynamic resistance can be estimated by the hum reduction method or by the voltage change over current change method. Typical values for test current and dynamic resistance are available from most catalogs.

If the extra lead is not objectionable, the signal tracer can be left attached to the zener during these tests. With practice, good zeners can be sorted from bad ones almost by ear alone, on the basis of hum and noise. But if this has not been done, a final check for noise should be carried out. Raucous, splattering noise indicates immediate disposal of the zener. A fine-textured hiss at low current levels is permissible, unless it shows a tendency to increase with time or current. Larger zeners should be firmly rapped with an insulating rod to check for loose internal connections.

Zeners that have passed all tests might be marked with fast-drying modeling paint, in resistor color code, as to their values. The paint will also serve to indicate that they have passed a fairly comprehensive test.

COMPUTER POWER SUPPLY TESTING

The hobbyist is frequently lured by advertisements for surplus power supplies and power supply kits of all kinds. These offerings range from excellent buys of surplus name-brand units to units, both kit and assembled, of unknown pedigree.

Performance specifications of these units sometimes are not fully stated in the advertising, but name-brand units can sometimes be researched in the manufacturer's literature. The "orphans" can literally be almost anything, and some surplus dealers are, at best, only brokers. There dealers may not have the necessary equipment, time, or interest to characterize surplus units they buy. Power supply kits may or may not have been carefully designed.

Where does this leave the hobbyist? Except for his purchase of name-brand units, he can be on a very shaky limb. Problems which might hit him can be one of three: complete failure, regulation failure, or the units may generate random and potentially harmful transients.

A complete failure is a nuisance at best and could be a costly disaster. The CPU can simply be shut down when the power supply quits, and few microprocessor systems have power -fail programming to save the run. Here are two possible disaster scenarios: a series-pass transistor shorts and dumps raw DC into the microprocessor systems, destroying all or most of the parts. Or, some critical power supply part fails, starts a fire, and destroys the house. Although the latter is unlikely, even commercial computers have been known to catch fire.

Regulation failures can be explained as a condition where the regulator circuitry ceases to operate in its linear region, usually because the input voltage to the series-pass element gets too low. Another problem can be thermal runaway if the unit has not been properly designed for continuous operation. A properly designed supply will protect itself and the load in case of shorts on the output.

The symptoms of regulator failure are low and unsteady output voltage, usually with high-ripple feedthrough. The effect on the CPU depends upon which logic family is used in the system. CMOS will be least affected, followed by MOS microprocessor and support chips, then TTL will be most affected. This nonlinear regulator will also have higher-than-normal output impedance, which leads to coupling between fast logic elements. The microprocessor system may drop an occasional bit, or become completely confused and lock up. Either way, the CPU is out of business.

Transients on the output come from the power line or are generated by an improperly designed power supply. Power line transients can usually be stopped by transient limiters on the power supply input, and output spikes generated by the power supply can be caught by crowbar circuitry. But, if the crowbar has to fire, the CPU will still be shut down in midstride. Properly designed power supplies will not generate spikes under any turn-on or turn-off condition. Of course, we all know that big spikes will kill any IC; we need to remember that small ones may cause improper system startup. Few home-brew systems are designed and programmed for standby operation, but main supply shutdown spikes could also bollix the remaining active circuits.

Fortunately, power supplies can be screened for most of these possible defects, using relatively simple equipment.

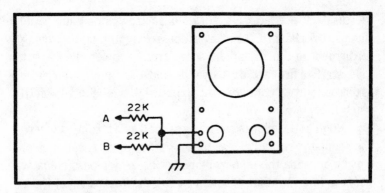

Fig. 1-61. Scope hookup for mixing two signals. Effective voltage of each signal is reduced to half.

When investments of the magnitude required to own a home processor are concerned, it pays to screen all power supplies not furnished by the system manufacturer. If any home system manufacturer should be guilty of supplying inadequate power supplies, it is likely that the computer grapevine will get the word out promptly unless he corrects the problem. For those hobbyists who design their own units, remember that solid-state power design is tricky, and even experienced designers test their prototypes thoroughly.

A modest amount of test equipment and two easily built gizmos will enable very extensive qualitative power supply testing, accurate enough to assure the hobbyist of adequate and reliable performance from power supplies he uses. *Quantitative* testing, required mostly to prove one's prowess at power supply design, will require more and higher quality equipment. The test equipment needed is the following: a 20 kilohms per volt VOM, a variable autotransformer, a pulse generator, a very stable power supply, and an oscilloscope. The scope should be dual-trace with triggered, calibrated sweep, 5 MHz bandwidth, and 10 mV sensitivity. The sensitivity and bandwidth are essential, and the triggered sweep will reduce frustrations during much of the testing. However, if a single-trace scope with the other qualifications is available, it is possible to get by with a couple of smart-aleck tricks. Figure 1-61 shows how two signals can be mixed on one trace. This trick is good only on two low-impedance sources, and it reduces the effective voltage of each signal to half its real value. In place of the calibrated sweep, let signal A be the signal to be timed, and signal B can be narrow pulses from a pulse generator with 1 kHw output. The resulting display will show 1 millisecond tick marks superimposed on signal A. If the scope has a Z-axis modulation input, the pulses can be used to intensify the trace as time markers.

The power supply will be used as a comparison standard to check regulation and stability of the power supply under test (PSUT), so it must be much more stable than the PSUT. If there is the slightest doubt, Fig. 1-62 shows a circuit suitable for a reference supply. All parts inside the dashed line should be built inside a thermally insulated box, and the supply should be allowed to run continuously for best stability. The output will vary from below 5V to around 18V, which should cover all supplies of interest to the computer hobbyist. The other equipment is very common, and the applications discussed in the following paragraphs will explain their use.

The first special gizmo is a dynamic load: a circuit which simulates the type of load fast logic places on a power supply. Figure 1-63 shows the circuit; R1 sets the load current drawn from the power supply for any given value of R2. Resistor R2 is chosen to give about 1V output at the maximum load current to be drawn from the PSUT. Capacitor C1 is an input from a pulse generator which causes the load current to increase sharply, then decrease, varying over a small range. Because Q2 and R2 dissipate the entire power output of the PSUT, both should be mounted on a large heat sink.

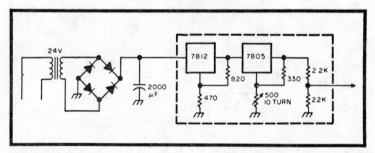

Fig. 1-62. Schematic of a suitable reference supply for comparing the output of any power supply under test.

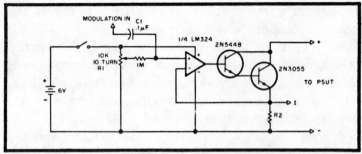

Fig. 1-63. Schematic of a special gizmo to simulate the load fast logic places on a power supply.

The second gizmo allows a power supply to be switched on or off at a special time: either at line-voltage zero crossing or at line-voltage peak. If the PSUT will develop spikes at turn-on or turn-off times, close examination of the PSUT output using an oscilloscope will reveal the problem. Figure 1-64 shows the circuit; it is safer to observe the special precaution of using an isolation transformer. If such a transformer is not available, take strict precautions to insure, by measuring both circuit common and point *L* to a good earth ground, that circuit common is really the low side of the line. Circuit operation is simple. IC1 samples the transformer output and produces approximately a square wave. With switch SW1 in the 0° position, the rising edge of the waveform at *B* will occur at zero crossing. With SW1 in the 90° position,

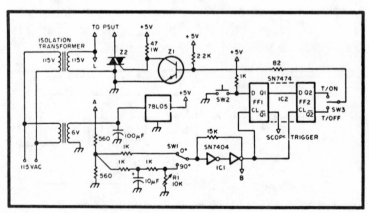

Fig. 1-64. Schematic of another gizmo that permits the power supply under test to be switched on and off at power line zero crossings or line peaks.

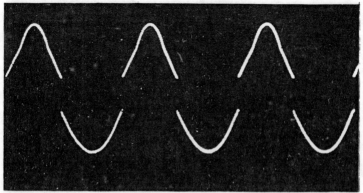

Fig. 1-65. Mixed waveforms from a circuit such as shown in Fig. 1-61 showing trigger at zero crossing of the line voltage.

resistor R1 can be adjusted so the trigger occurs at line voltage peak. The waveforms in Figs. 1-65 and 1-66 show the mixed waveforms (see Fig. 1-62) for zero crossing (Fig. 1-65) and for switching at line peak (Fig. 1-66). IC2 consists of two flip-flops which are clocked by the output from IC1. If switch SW2 is open, Q1 and Q2 are both high; with SW3 in the T/ON position, Z1 is turned on and triac Z2 is turned off. When SW2 is closed, Q1 goes low on the next positive edge from IC1, and Q1 triggers the scope. On the second positive edge from IC1, Q2 goes low and Z1 turns off. Z2 is then turned on by current through the 47Ω resistor, and any load connected to Z2 is activated.

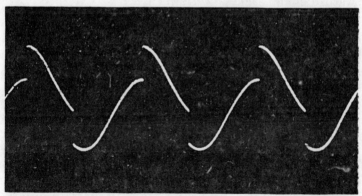

Fig. 1-66. Mixed waveforms showing proper adjustment of pot R1 of Fig. 1-64 so that IC1 triggers at line voltage peak.

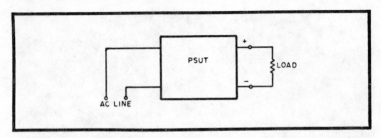

Fig. 1-67. Typical connection for a fixed voltage power supply with no sense lines.

After R1 has been adjusted properly, the load can be switched at zero crossing or line peak by setting SW1 and closing SW2. With SW3 in the T/OFF position, Z2's load is activated until SW2 is closed; then Z2 turns off at the next zero crossing after FF2 switches.

Three types of tests should be performed on all unknown power supplies to give reasonable assurance of safe, reliable operation in a home computer system. These are: (1) transient detection and rate of rise, (2) DC regulation and heat rise, and (3) output impedance.

As mentioned above, power supply transients can disrupt or damage the CPU. In a computer system which uses several different voltages, sections of the system may lock up or be damaged if, for example, the +15V bus comes up before the −5V bus. The rate of rise measurement allows the system designer to know if all supplies will come up together when switched by the same power switch.

The DC regulation test tells how well the regulator circuitry performs, and the heat rise test forewarns of possible fire or overheat conditions if the supply is used for continuous operation.

A test of power supply impedance tells how well the supply will decouple possible circuit interactions. A noisy Vcc line can be very bad for TTL and may cause subtle problems almost impossible to locate without a very good oscilloscope. Although decoupling on the board is helpful, it will not entirely overcome the effect of a high-impedance power supply. Probably 90% of all decoupling would not be needed if a perfect (zero output impedance) supply were available; *however*, a

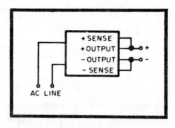

Fig. 1-68. Power supply sense lead connections of a power supply under test.

perfect power supply feeding a load with 2 feet AWG #24 clip leads might have over an ohm of impedance for fast logic signals.

Figure 1-67 shows typical output connections for a modular, fixed voltage supply which does not have sense lines. Any stated performance specifications (percentage of regulation, noise, ripple, etc.) apply *only* at those terminals. They will still be very close if the supply feeds a load via AWG #8 leads. If the supply is feeding a home-brew backplane wire wrapped with AWG #30, the top of the card 12 slots away may be in trouble!

Some of the connecting lead problems go away if the supply has sense leads as shown in Figs. 1-68 and 1-69. Sense leads must connect electrically as shown in Fig. 1-68. If this connection is made physically at the supply, the equivalent circuit is the same as Fig. 1-67. If both the sense leads and load leads attach at the load as shown in Fig. 1-69 the published specs now apply at the load.

Power supply test hookups must be carefully and properly made. Figure 1-70 shows right and wrong connections; note that the wrong connection includes the effect of the load leads and the drop across the ammeter in the measurement. If you want to prove the importance of heavy power supply

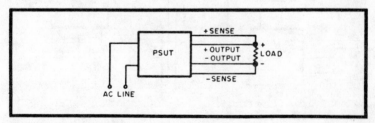

Fig. 1-69. Connecting the power supply sense leads across the load to obtain accurate adjustments to the output.

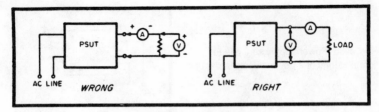

Fig. 1-70. Right and wrong method of test hookups.

leads, repeat some of the tests outlined below using the wrong connection and AWG #30 wire and check the difference.

Do the transient test first and discard or redesign any unit which fails. Use the triac switch of Fig. 1-64 with the connections shown in Fig. 1-71. *Remember the grounding precautions if an isolation transformer is not used.* With SW3 set to T/ON and SW1 set to 0°, sync the scope so the rising output of the PSUT is displayed on the scope as in Figs. 1-72 and 1-73. Repeat the test for T/OFF (Fig. 1-75). and repeat the T/ON test with SW1 set to 90 degree (Fig. 1-74). Okay the PSUT only if no spiking occurs under low, medium, and full load conditions. If rate of rise is important to your system, measure it before dismantling the test setup. The PSUT waveforms shown in Figs. 1-72, 1-73, 1-74, and 1-75 are from a surplus supply with a good pedigree; it is very gratifying to have such excellent test results from a supply which cost about half the going rate.

Set up the regulation test as shown in Fig. 1-76. Select R2 of the dynamic load (Fig. 1-63) for 1V drop at full rated load

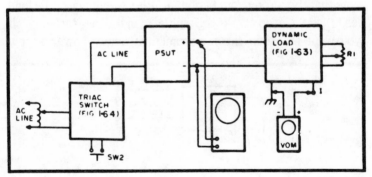

Fig. 1-71. Test hookup for transient detection. Use the triac switch of Fig. 1-64 and the dynamic load in Fig. 1-63 in this hookup.

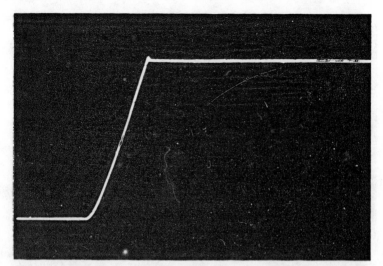

Fig. 1-72. Scope waveform of a 5V 3A supply at turn-on time. The supply has a 1.8A load switched at line-voltage zero crossing. Rise time is 3 milliseconds. Note the rapid response with very little overshoot.

current of the PSUT, then read current with the VOM at output I. For example, if R2 is 1Ω, 1A will give 1V across R2. Set the line voltage to 115V, and set the reference supply to the same output voltage as the PSUT. Let the whole system warm up at 50% load (PSUT rating), then connect the VOM

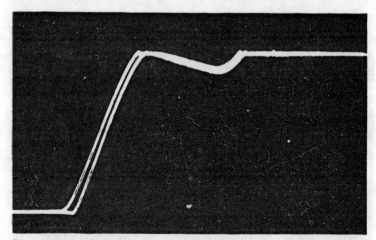

Fig. 1-73. Waveform double exposure of the 5V 3A supply at turn-on time. The same 1.8A load is used and switched at zero crossing. Voltage irregularity caused by incomplete charging of filter capacitor on first cycle of line voltage. The voltage does not exceed the set value by more than 1%.

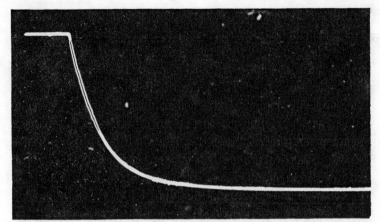

Fig. 1-74. Waveform of the 5V 3A supply at turn-off time with a 1.8A load. Total decay time is 150 milliseconds. Note the complete absence of transients.

with the resistive network shown in Fig. 1-76. Set the VOM to the 50 μA scale and adjust the reference supply so the meter reads zero. Close SW4 and readjust the reference supply for midscale on the meter. Now, 50 mV change in PSUT output will swing the meter to end scale. Lower the line voltage to 105V; note the VOM reading after 1 minute; then raise the line to 125V and note the change after 1 minute.

Percentage regulation is defined as

$$\frac{V_H - V_L}{V_H} \times 100\%$$

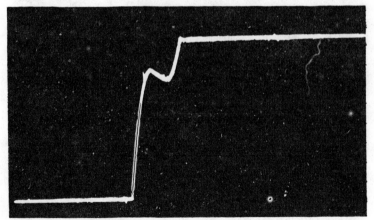

Fig. 1-75. Waveform of turn-on time of the same 5V 3A supply with a 1.8A load. This time turn-on time is at line peaks.

where V_H is PSUT output at high line and V_L is PSUT output at low line. Since this connection will measure the change in output, this reduces to the VOM reading in millivolts divided by the output voltage set on PSUT (an approximation, unless a digital voltmeter is available to measure V_H). For 100 mV change (full scale swing in the VOM with SW4 closed) on a 5V supply, the percentage of regulation is 0.1/5, or 2%. This is not good unless the impedance test gives exceptionally good results. Even so, such poor regulation on a supply sold as a regulated supply may mean the regulator circuit has problems.

Repeat the test with line voltage fixed at 115V, and change the load from no load to full load. Compute the regulation as before, using the change in output voltage as measured by the VOM and the nominal output voltage of the PSUT. Acceptable regulation is 0.2% (10 mV change on a 5V supply), but 0.1% is better for most applications. For that supply with poor line regulation, repeat the no load/full load test at low line voltage, and reject it for long-term use if the regulation changes substantially at low line voltage.

Run the heat test by pulling full load for several hours. The PSUT output should change very little, and no part of the supply should become too hot to touch. If the heat rise is too much or the regulation changes drastically, reserve the unit for smaller loads or intermittent duty. Sometimes it is possible to add heat sinks to overheating supplies, but heat-flow design is beyond the scope of this book.

The output impedance test (Fig. 1-77) also uses the variable load. In this case, set some midrange load on the

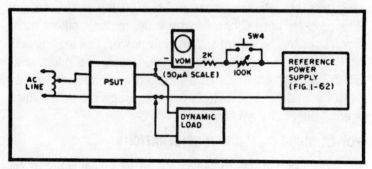

Fig. 1-76. Setup for regulation test of power supplies.

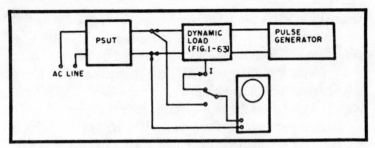

Fig. 1-77. Setup for impedance test.

PSUT, then feed a small amplitude square wave or pulse to the modulation input. Use the scope to monitor the *I* terminal on the variable load, and set the modulation drive so the current is changed by 10% of the fixed load current. Then move the scope to the PSUT output and look for output voltage changes at the modulation frequency. Measure both the current change and output change in voltage peak-to-peak and divide the voltage change by the current change. Output impedance is simply this voltage divided by the current. For example, assume the load current is 1A, with a current change of 0.1V peak-to-peak. If the PSUT output changes by 0.1V peak-to-peak, the impedance is 0.01/0.1, or 0.1Ω, a good value. Vary the pulse frequency over a fairly wide range to check for any major change in impedance. Most solid-state supplies of good design will hold output impedance essentially constant over a wide frequency range.

Now that you have located a good supply with low impedance, don't handicap it. Run bus *bars* for power lines instead of wires, solder power connections everywhere except where they pass through connectors, use high quality connectors and choose or lay out PC boards with wide, multiple power runs. Even if the logic card draws fairly low power, fast logic generates fast transients. Small power conductors are inductive at much lower frequencies than large, flat conductors. Inductance in a power bus encourages interstage coupling, which spells *trouble* in any computer.

POWER SUPPLY DESIGN CONSIDERATIONS

The first part of any power supply to look at is the power transformer, diode rectifier, and filter combination. What is desired is to get a transformer with a rating of greater than 12V since what is desired is a regulated output of 12V to 14V. It would be best to get a transformer of about 15V RMS rating, but this is not a very common transformer. A transformer of a higher voltage rating (20V or so) is okay, but this would mean the regulator would have to dissipate a lot of heat, which wastes power.

The best compromise is get a transformer which will deliver a solid 12.6V at 3A (See Fig. 1-78). Some of the less expensive transformers will drop several volts at full current; try to avoid these. You may wonder how to get a regulated 12.6V from a 12.6V transformer. The trick is in the rating of the transformer in RMS voltage.

Since AC voltage is constantly changing in value, at any given instant, its absolute voltage may be anything from zero to the peak value. Because of this, AC voltage is given based on how much work it will do. A potential of 12.6V AC RMS will do the same work (heating a resistor) as 12.6V DC; however, the peak AC voltage is 1.414 times the RMS voltage. A simple rule of thumb is, split RMS in two, and add this figure to the RMS. This would be say, 12V divided by 2 equals 6; add this to 12V for 18V peak.

This method gives a ballpark figure which is close enough to work with. Using this figure of 18V, rectifier diodes which have a reverse breakdown rating of greater than 18V and can handle current in excess of 3A are required. This part is easy, since most power rectifier diodes handle at least 25V.

In selecting the diodes, it is best to allow a margin of safety to allow for current surges and voltage spikes. The

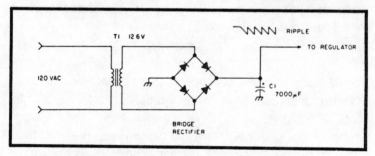

Fig. 1-78. Schematic of typical unregulated bridge rectifier circuit.

turn-on current surge could damage the diodes when the power supply is first turned on, but the resistance of the transformer's winding is usually enough to protect the diodes (which are able to handle currents on the order of 50 times the diode rating) for a very short period of time. In this case, a 5A, 60V bridge would be good.

Now comes the filter. There are several types of filters, but consider only the most common, which is the capacitive input type. This filter is simply a capacitor across the output of the rectifier. This capacitor will, under no load conditions, charge to the peak value of the voltage from the transformer, or, in this example, to about 18V. The capacitor must have a voltage rating of at least 18V, preferably 20V.

One of the major pitfalls is encountered in the selection of this capacitor. It does not need to be so large as to give pure DC at full load. Its value should only be large enough to prevent the voltage drop between charging pulses from the rectifier from dropping below the point where the regulator falls out of regulation.

Most regulators will require no more than 3V across the series pass transistor to maintain regulation. That being the case, there could be from 18V to 12V at the output of the bridge. With a 3V requirement on the pass transistor, there must be a minimum of 15V at this point before the circuit would drop out of regulation. Fifteen volts minimum from 18V maximum gives a maximum of 3V allowable ripple. To have a larger capacitor and less ripple serves no useful purpose and wastes space with the larger capacitor. To determine the value of this capacitor, figure the regulator impedance as seen from the filter capacitor under full load. Using Ohm's law under a full load of 3A at 18V peak,

$$R = 18V/3A$$
$$= 6\Omega$$

A simple review reminds us that t equals RC, or resistance (ohms) times capacitance (farads) equals time (seconds) for a 63% charge or discharge of a capacitor. If the discharge curve of a capacitor is drawn and the top peak is shown to be 18V, pointing out the 63% discharge point (6.6V), it can be seen what percentage of the discharge time it takes for the capacitor voltage to drop below 15V (regulated output voltage plus 3V).

The 3V allowable discharge is about 16% of the charge, which becomes 84% charge at about 0.2 time constant. A greater peak voltage allows more ripple to play with, hence a smaller filter capacitor. In this case it will require 5 (1/0.2) time constants to not drop below 84% of charge between charging pulses from the rectifier. The power supply will be running from 60 Hz AC and the full wave rectifier will put out charge pulses at a 120 Hz rate, or one pulse every 8.33 millisecond (ms). Five times this figure gives a required time constant of 0.0416 seconds. Now it is known that R equals 6Ω and t equals 0.0416 seconds, and that T/R, or 0.0416/6, equals 0.0069333 farads or 6933 μF. This says that with the transformer and current requirements given, this power supply needs 7000 μF for filtering.

After what has just been discussed, it should be mentioned that the regulator does not remain a constant 6Ω. This is because the regulator adjusts to maintain regulation and as such is more of a constant current sink. This being the case, the discharge curve will be a straight line (linear) until the regulator drops out of regulation. When dealing with the upper part of the discharge curve, the difference is unimportant, and falls into the area of slop to be covered by the margin factor.

Now for the regulator, another pitfall. Nowadays the easiest thing to do is use an IC regulator with a external pass

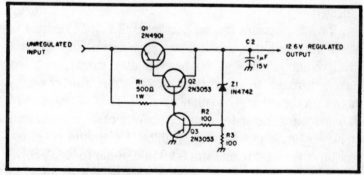

Fig. 1-79. Schematic of regulator circuitry.

transistor. The IC route won't be followed since we would not learn much about how it works.

Refer to Fig. 1-79 for the regulator circuit. The series pass transistor will be selected mostly by the current capability. What is desired is a PNP that can pass 3A and dissipate 15W or so. The dissipation required is determined by the voltage drop across the series pass transistor at full load. Since at that point the filter capacitor will also have maximum ripple, giving lower average power, a safety margin can be added by maintaining the full voltage differential for the calculations:

$$18V - 12.6V = 5.4V$$

therefore,

$$5.6V \times 3A = 16.2W$$

which is the worst-case dissipation. A small heat sink and almost any power transistor can handle that. To parallel series pass transistors is a total waste, in this case, as almost any power transistor can handle this power level.

To point out a common mistake, you cannot just parallel two transistors. Refer to Fig. 1-80A. It won't work, although it will appear to. The base emitter junction is electrically a diode with a conduction voltage of about 0.6V for a silicon transistor, and about 0.2V for a germanium transistor. It may be assumed that no two junctions are exactly alike. Because of this, one will conduct before the other, not allowing the base voltage to rise further to the conduction point of the second transistor. Thus it will not turn on unless the first transistor is damaged.

Two transistors can be paralleled if a small resistor is placed in the emitter circuit of each transistor. This allows a small voltage change in the base emitter circuit, with the current through the transistor. If one transistor conducts more than the other, the voltage builds across the junction of the conducting transistor, causing the other transistor to turn on harder, striking a balance. Without this emitter balancing resistor, the second transistor is wasted. Refer to Fig. 1-80B.

Now to pick a series pass transistor, for example an inexpensive 20W PNP with some gain. As you can see, almost

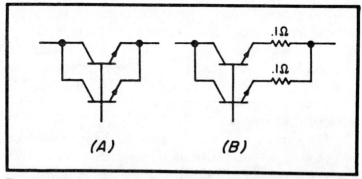

Fig. 1-80. Incorrect (A) and correct (B) methods of paralleling two transistors.

any PNP power transistor will do, but pick a 2N4901. This transistor is rated for 40V, 85W, and costs about $1.75 from Motorola. The gain for this device is a minimum of 25, which means that at 3A you must be able to supply a worst case base current of 120 mA. The best way to do this is to use another transistor which will amplify a much smaller current. For the driver transistor in this case, pick a 2N3053, which has a breakdown voltage of 40V and a minimum gain of 25. Referring to Fig. 1-79, wire Q1 and Q2 together, as shown, in a Darlington configuration.

Now to figure the value of resistor R1, which must supply enough base drive to provide the worst case drive to Q2. Since Q2 must drive Q1 with a maximum of 120 mA (output current divided by gain of Q1), derive that 120 mA by the gain of Q2 to get a rounded off 5 mA drive requirement for Q2. At 3A output, the filter capacitor voltage will be dipping to as low as 15V so with a worst case low input voltage of 15V and division by 0.005A, the resistor value should be about 500Ω.

To figure the power dissipation of the resistor, take the maximum voltage across it, which occurs at no load, with maximum charge on the filter capacitor. The no load voltage will be about 18V. The base of Q2 will always stay within about 0.5V of the output voltage, so the worst case voltage across R1 will be about 6V. Again, using Ohm's law, take the 500Ω and divide into the 6V to obtain current at 120 mA. I times E equals P, or 0.12A, times 6V equals 0.72W. A 1W resistor would be in order here.

Now for the feedback loop which makes this into a reg-ulator Z1 is a 1N4742, 1W 12V zener diode. It is connected as shown with R3, whose only purpose is to sink any leakage the diode might have and to assure that the diode operates in its proper current range. The value is not critical and the 100Ω shown will allow about 6 mA to flow through the diode before regulation takes place.

The purpose of R2 is only to protect Q3 against a sudden current surge that might damage the base junction before the regulator can respond. With Q3 installed as shown (almost any NPN transistor will work), any current in excess of the 6 mA designed to flow through Z1 will turn on Q3, causing it to steal base current from Q2 until the output voltage reaches the cutoff point of Z1. The circuit won't allow the output to go above 12.6V (the extra 0.6V comes from the base junction of Q3), and if it tries to go below 12.6V, Q1 and Q2 drive harder and won't allow that. So what we have is a rock solid 12.6V.

Now for the final pitfall. C2 is not a filter capacitor. To try and filter at this point will waste a big capacitor. C2 should only be a small capacitor to *bypass* any high frequency noise which might appear at this point. A 1 μF will do just fine.

In conclusion, while this is intended to show how to design a simple power supply and avoid some of the common pitfalls, if built it will provide a solid, pure 12.6V at a continuous 3A. In fact, since all was figured on worst cases, it will supply in excess of 3A and not feel much strain. Current limiting and overvoltage protection could be added with little effort.

MORE THOUGHTS ON PS DESIGN

The power supply is usually the very last part of the circuit to be built into an electronic design. This is perhaps as it should be, since the circuit designer often doesn't *really* know the exact voltage and current requirements until the design is done. But, because the power supply *is* last, it often is a victim of the dollar-short-and day-late effect. Also, the power supply may suffer at the hands of a designer too well steeped in the AC/DC broadcast receiver practice, to the point where power supply and a rectifier and a capacitor are synonymous.

The power-supply should not be an afterthought, but rather should be well designed; since a good power supply can effect many simplifications in the associated circuit design. Also, in this age of very inexpensive diodes and transistors, the difference between a rough power supply and a nice power supply is only a matter of a couple dollars worth of parts.

If one accepts the idea that a good power supply design is desirable, will result in a better overall circuit, and will save design time to boot, how does he approach the total design? If one first builds himself a laboratory power supply which has *all* the features of any supply he could want, he can use it to try out each circuit, as it is designed. Then, having a circuit that works well with the lab power supply, one can design a simpler power supply that encorporates only the *needed* aspects of the laboratory unit.

Transformers Can Be Inexpensive

The design of a power supply almost invariably requires the use of a transformer. This is for two reasons: isolation from the power line and obtaining a voltage different from the line voltage. The transformer is likely to be the highest priced single item in the power supply, so resourcefulness in choosing a power transformer can really save money. Utilization of transformers that are available everywhere avoids buying the special types with their high price tags.

It is almost a truism that any electronic part we can use that is made for an automobile or a TV set is at the rock-bottom price for an item of its type. This is simply because the TV set and the auto represent essentials to the American public—everyone has at least one. Further, junk TVs and autos abound. The large number of amateur transmitters powered by old TV set transformers is witness to this fact. One can find 6.3V transformers in some of the junk TV sets; they were used to provide heater voltage to some or all of the tubes, while plate voltage was obtained using a semiconductor doubled directly from the power line. These 6.3V heater transformers can be useful in transistor power supplies employing the full-wave bridge or conventional doubler circuits.

The early 12V auto radios (with tubes that utilized + 100V to + 150V for plate voltage) are another source to a power transformer. If we simply connect the 117V AC line across half the HV secondary (center tap to one side), we will get about 12V AC each side of the primary centertap.

A surplus 24V vibrator transformer represents an even more useful find. If the HV secondary is about 100V each side of center tap, such a transformer will put out about 24V each side of the primary center tap. This higher voltage is more desirable in most instances.

Oh yes, don't forget to dig the silicon or germanium rectifiers out of the old TV set; they work just as well for low-voltage transistor power supplies as they did supplying several hundred milliamperes of B+ to the TV tubes. Although large by modern standards, Sarkes Tarzian M500's and the like have worked very well in the low-voltage power supplies.

Rectifier Circuits

In the area of 60 Hz single-phase rectifiers, there are five types of circuits that we'll be concerned with. These are half-wave, full-wave, bridge, conventional doubler, and cascade doubler. These are illustrated in Fig. 1-81.

The full-wave, bridge, and conventional doubler circuits all charge their filter capacitors continuously throughout the 60 Hz period. This continuous type of rectifier output waveform contains no 60 Hz components, but only 120 Hz, which is more easily filtered. The half-wave and cascade doubler circuits, having considerable 60 Hz energy in their output waveforms, will prove to be the most difficult to filter and will have poorer regulation than the others. However, these relatively less desirable circuits are the only circuits one can use if one side of the transformer secondary *must* be grounded for some reason.

With all these rectifier circuits at our disposal, *and* with the additional variation of being able to choose either capacitor input or choke input to our filter section in full-wave or bridge circuits, a given transformer can provide us a variety of DC output voltages.

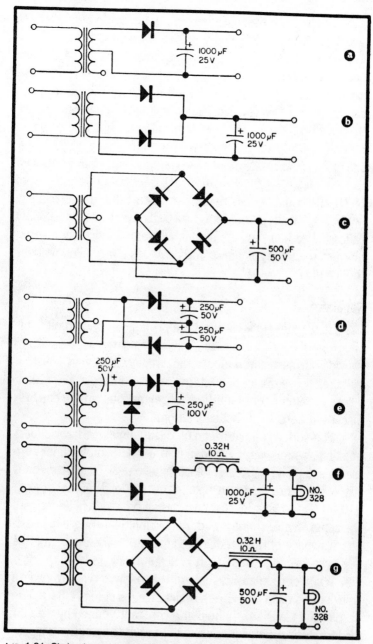

Fig. 1-81. Six basic types of rectifier circuits. (A) Simple half-wave rectifier. (B) Full-wave rectifier. (C) Bridge type. (D) Conventional full-wave doubler. (E) Cascade half-wave doubler. (F) Full-wave rectifier with choke input. (G) Bridge rectifier with choke input. Performance of these circuits is shown in Fig. 1-82.

To illustrate the variations of performance available with one transformer, a Triad F40X 24V center-tapped transformer was used in all the seven variations of Fig. 1-81. Then each circuit's voltage-current characteristic was tested; these are shown in Fig. 1-82. To make the circuits comparable, the capacitor bulk (microfarads times voltage) was kept constant throughout. That is, the constant bulk concept would equate 500 μF at 50V with 1000 μF at 25V. Note that in Figs. 1-82F and 1-82G (the choke input cases) that a pilot lamp was added as a bleeder resistor. For full-wave rectifiers operating from 60 Hz, R should be less than or equal to 1130L (where L is in henries). For our example, L was 0.32 H so R should be 362Ω or less. The number 327 bulb is approximately such a resistance at the voltages used, and provides a free pilot lamp for the supply.

Regulators

Having developed a number of rectifier-filter circuits, it is clear that they *all* have some lack of regulation. If we wish to provide our associated electronic circuit with a very constant voltage, some form of voltage regulation must be added. There are a number of methods of regulation, but the one first discussed here is series regulation.

The series regulator can be easily developed, for understanding, as follows. Consider first of all a simple zener regulator as in Fig. 1-83A. The output voltage characteristic is that of the zener diode, providing that the load current doesn't exceed (Es − Ez)/R. A capacitor can be added in shunt with the zener diode, to afford additonal ripple filtering. One serious drawback of the zener regulator is that the zener voltage does change somewhat with zener current.

A better regulator is the emitter-follower type, wherein a transistor is used as a series resistance, as in Fig. 1-83B. The base of the transistor is held at a constant voltage by a zener diode, which derives its current through R from the unregulated input. Since the current flowing into the base is Ic/hFE, the zener current can be much less than in the simple zener regulator. If we make R small so that the zener current is large

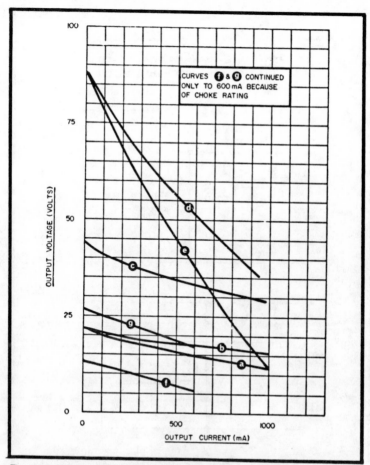

Fig. 1-82. Performance of the rectifier circuits shown in Fig. 1-81.

compared to the maximum base current the transistor will ever draw, the percent variation in zener voltage can be made fairly small. Again, a capacitor can be put in shunt with the zener, giving the ripple reduction of the well known capacitive multiplier.

Perhaps the next step, in sophistication, is to add DC gain to our series regulator as in Fig. 1-83C. In this circuit one is not tied down to the zener voltage to determine the output voltage. In fact, the output voltage is adjustable, over a few percent, with the output sample-pot, Rs. The circuit still suffers somewhat from variations in zener current with load current, however.

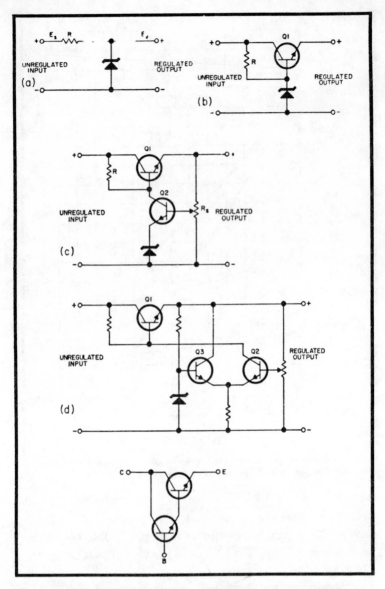

Fig. 1-83. Regulator circuits. (A) Zener regulator. (B) An emitter-follower regulator. (C) An emitter-follower regulator with amplifier. (D) An emitter-follower type with full differential amplifier. (E) Darlington pair that acts as a single transistor with higher gain.

Finally, then, we add the full differential amplifier for our gain stage as in Fig. 1-83D. Now, since the zener diode derives its current from the output side of the regulator, the zener current is nearly constant and so is the zener voltage. As in Fig. 1-83C, the output is adjustable over a few percent.

In all the circuits of Figs. 1-83B, 1-83C, and 1-83D, it is possible to replace series transistor Q1 with a Darlington pair; that is, replace one transistor with two as in Fig. 1-83E. This combination yields the equivalent of a transistor with the product of the hFE of the two component transistors. This will allow us to use higher DC gain in the regulator system, affording better performance. The Darlington pair method is a way of obtaining a high hFE power transistor for use as Q1.

Examples of Practical Circuits

Having looked at a variety of series regulators, a selection of real-life examples to illustrate them may be in order. These are presented in Figs. 1-84, 1-85, and 1-86. The circuits are not so designed to make value judgments between types, but are simply examples.

Another differential regulated supply is presented in Fig. 1-87 to illustrate a trick in substituting the base-emitter junctions of inexpensive silicon transistors for zener diodes. Note that in both Figs. 1-86 and 1-87 a preregulator zener diode is used to supply half the differential amplifier with collector voltage.

One will also notice that in the examples of Figs. 1-84, 1-85, 1-86, and 1-87 that there is added, across the transformer secondary, a 0.01 μF 1 kV disc ceramic capacitor. This is a line transient suppressor whose purpose is to kill any line

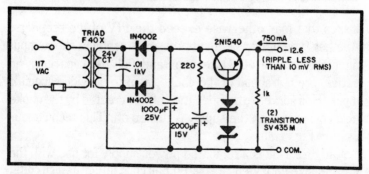

Fig. 1-84. Simple emitter-follower regulator that was used for the heaters of five 12AX7s.

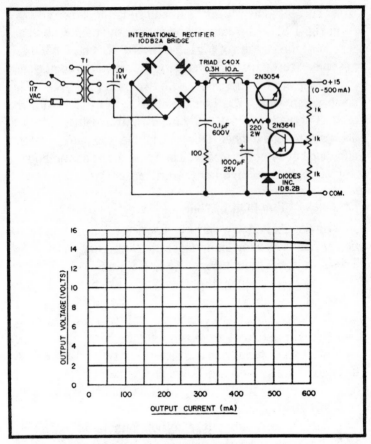

Fig. 1-85. Circuit and performance curve of a regulated power supply using the basic circuit of Fig. 1-83C. Transformer T1 is a surplus vibrator type rated at 25.2V input and 135V 118mA output with a vibrator frequency of 115 Hz. Half of the high voltage winding is used in this circuit.

spikes that may otherwise exceed the PIV of the rectifiers. Another type of transient suppressor is useful in choke-input filter systems; a 100Ω resistor and 0.1 μF capacitor, (in series) are put ahead of the choke, as if to provide capacitor input. This serves to dampen the voltage created by the choke field collapse when the supply is turned off. This technique is used in Fig. 1-85.

Figure 1-88 is a regulated power supply for use with the inexpensive epoxy-encapsulated Fairchild integrated circuits. These epoxy ICs are specified for +3.5V to −4.5V. This regulator utilizes one of the old 2.5V filament transformers that was put in the junkbox when you replaced the 816s or 866As with silicon HV rectifiers. The Fairchild μL923 (a J-K flip flop) takes about 20 mA, so this power supply will run up to fifteen of them.

This design, a low-voltage one, brings out several of the worst points in series-regulator design. Since the difference between rectifier voltage (point A) and the regulator output voltage (point C) is only a few volts at most, the (small) voltage drop of the base-emitter junction becomes appreciable percentagewise. Since germanium transistors have lower emitter-base drop, a germanium unit is used here; a high hFE germanium type was chosen, the 2N2147.

Figure 1-88 is similar to the regulator of Fig. 1-83B except that R has been replaced by a #49 (2V, 60 mA) pilot

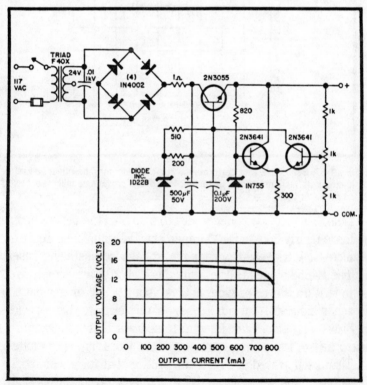

Fig. 1-86. Schematic and performance curve of a practical supply using a differential pair amplifier.

111

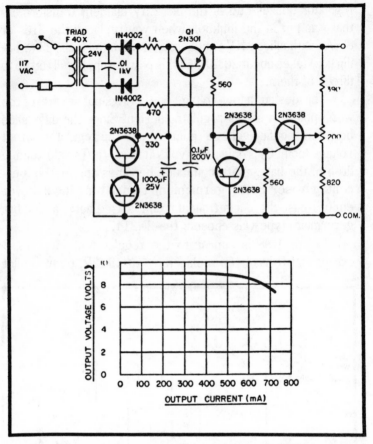

Fig. 1-87. Schematic and performance curve of another differential pair regulated supply. Here the base-emitter junctions of inexpensive transistors are used as zener diodes.

lamp. This trick was necessary to help hold the zener current more nearly constant. The bulb acts as a nonlinear resistor, increasing its resistance as the voltage across it increases. The 1N658 diode in series with the 1N4728A was simply used to jack up the reference voltage about 0.7V, over what the zener alone turned out to be. In this circuit, this forward-biased diode *does not* effect a temperature compensation.

Also, low voltage brings out the worst in zener diodes. This is illustrated in Fig. 1-89. Notice that for zener diodes with voltages below about 6V, the "knees" do not have a sharp break, but are quite rounded. This phenomenon is associated

112

with the different mechanism of breakdown at the lower voltages. (To be really correct, one should only call regulator diodes that break down below about 6V "zeners" since this is in the zener region. Similarly regulator diodes that break down above 6V should be called "avalanche" diodes. However, the technology has come to call them *all* zeners.) The "soft knee," as this rounding over at breakdown is called, can be coped with by running more zener diode current within the dissipation limits of the diode.

This soft-knee behavior of low-voltage zeners is not all as bad as it might be. It turns out that the zener and avalanche regions of breakdown have opposite temperature characteris-

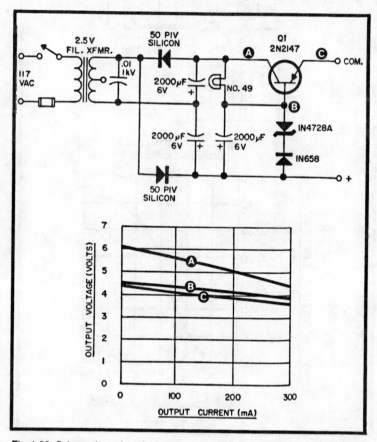

Fig. 1-88. Schematic and performance curve of a regulated supply for use with inexpensive Fairchild ICs. The letters on the graph refer to voltages at the points shown on the schematic.

tics. So that at about 5V (see Fig. 1-89B) the two opposite temperature coefficients cancel, yielding a ready-made temperature-compensated reference diode.

At breakdown voltages above 5V or 6V, we are operating with a positive temperature coefficient, in the avalanche reg-

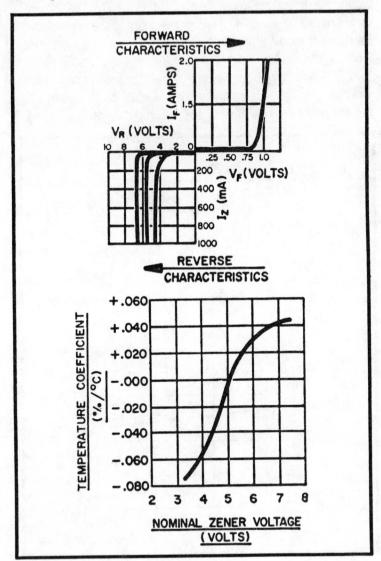

Fig. 1-89. Zener curves. (A) The reverse characteristics of low-voltage zeners. (B) How temperature coefficient of zeners is lowest at about 5V.

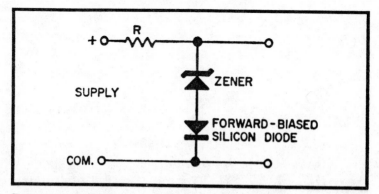

Fig. 1-90. Temperature compensation of a zener diode when used with types greater than 5V or 6V.

ion. Since the temperature coefficient of an ordinary silicon diode or rectifier, when forward biased, is negative, the arrangement of Fig. 1-90 can be used to help temperature-compensate higher voltage breakdown diodes.

Chapter 2
Regulated Low-Voltage Power Supplies

This chapter probably holds the most interest for the hobbyist since it contains the type of circuits needed by today's solid-state circuitry. You'll find a wide range of regulated supplies here, starting with types that regulate down to 0.35V and up to 50V. After 50V the term low voltage must be changed to high voltage or medium voltage, which really isn't defined. But for our purposes here we'll call everything under 50V a low-voltage type; anything over that figure is a high-voltage type.

12V 20A BRUTE SUPPLY

This power supply (Fig. 2-1) is intended to deliver 12V to 15V at about 20A intermittently and 9A to 11A continously. You should have no trouble with the circuit except possibly locating some of the parts.

All of the parts on the regulator board are very easy to find if you use ½W resistors. Radio Shack and Lafayette seem to keep an ample supply. The same with the LM723 chip.

Good high current pass transistors will work well here. The best you can get is the 2N3055 or equivalent, as they are less susceptible to going haywire with some RF feedback. You can drive four of these goodies with a single 2N4911 or equivalent and, by putting 0.3Ω resistors on the emitters of

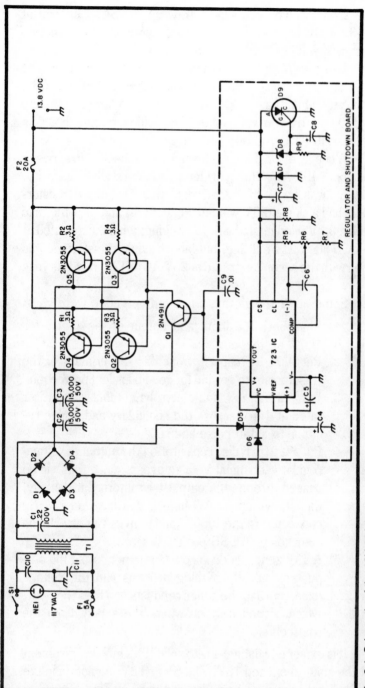

Fig. 2-1. Schematic of the 12V 20A brute supply.

117

the 2N3055s, you can safely draw up to 25A of current at +13.8V DC intermittent and about 18A continuous, that is, provided you heat sink all of the transistors.

Use two 15,000 μF capacitors for better regulation, but they don't have to be two of that exact value. You can use what you can find and, as long as they total up to between 25,000 and 30,000 μF, you will be okay. The voltage ratings should be at least 40V DC and 50V DC surge.

A couple of old TV transformers were hooked in series to get about 19V AC. With a bridge rectifier and all the capacity, 25V AC was present for good regulation. One of the transformers had a single 6.3V winding at about 8A and the other had both a 6.3V winding and a 5.0V winding, which were rated at 7.5A. This works out fine, but later a single transformer was used with the secondary rated at 25V AC at 12A; it works very nicely.

Some other things that improved operation were:

1. Fuse both the input and output of the supply for 100% protection.
2. Put 0.01 μF capacitors across the primary of the transformer to ground for good transient protection.
3. Put a 0.22 μF non-electrolytic tubular capacitor rated at 100V across the secondary and bridge rectifier to aid in noise and rf.
4. Put a 0.01 μF disc capacitor on the output of voltage regulator terminal V$_{OUT}$ to bypass noise that will be passed through the output transistors. Incidentally, this shows up as AC hum if you have an FM rig connected to the supply, and it gives you an echo on your voice on SSB peaks.
5. A 15V zener diode is constantly monitoring the output voltage. If anything happens and the voltage exceeds 15V, the zener conducts and fires the SCR, which immediately crowbars the supply from the transceiver.

It is extremely dangerous to solid state mobile equipment for the input to exceed 15V. The SV and 8V regulators in the rigs will not take it. A parts list is shown in Fig. 2-2.

Parts List

R1-R4	.3 Ohm 10 Watt
R5	1.8k ½ Watt 10%
R6	2.5k trimpot
R7	2.7k ½ Watt 10%
R8	1.5k ½ Watt 10%
R9	1k ½ Watt 10%
C2-C3	15,000 uF 40 V dc
C1	.22 uF @100 V dc tubular
C4	250 uF @25 V dc
C5-C6	1.2 uF @35 V dc tantalum
C7	220 pF @25 V dc
C8	100 uF @25 V dc
C9-C11	.01 uF @500 volts
D1-D4	1N3492 or equiv. (100 piv @ 18 Amps)
D5-D6	1N4607 or equiv.
D7	1N4002 or equiv.
D8	1N965A or equiv. (15 volt zener)
D9	2N4441 or equiv.
Q1	2N4911 or equiv.
Q2-Q5	2N3055 or equiv.
F1	5 Amp fast blow fuse @ 125 volts
F2	20 Amp fast below fuse @ 125 volts
NE1	115 V ac neon indicator lamp
S1	Switch — single pole, single throw @ 115 V ac

Fig. 2-2. Parts list for the 12V 20A brute.

VOLTAGE DOUBLER FOR MOS/LSI

Figure 2-3 shows two common voltage doubler circuits. In Fig. 2-3A is the conventional full-wave doubler, and in Fig. 2-3B is the cascade or half-wave doubler cirucit. Each of these circuits offers the same advantages or disadvantages as its namesake in the standard rectifier circuit. The full-wave doubler offers higher average output in both current and voltage, depending on the load applied to it, and requires less filtering. The half-wave circuit uses fewer parts but does require much heavier filtering to achieve a similar output.

For the needs in this application, the half-wave circuit has one major feature which dictates its use. It has a common input and output side that can be grounded for a reference to the main power supply.

To understand how a voltage doubler operates, please consider Fig. 2-3B. The voltage of this supply is taken across capacitor C2, which is charged on alternate half cycles. When the input is in its negative half cycle, D2 conducts and charges capacitor C1 to the peak input voltage. On the next half cycle of input, the input signal reverses polarity and is now in series with the voltage across C1. The voltage at point A is now equal to twice the peak input voltage. Diode D1 now conducts and charges capacitor C2 to this higher voltage. The output voltage of the doubler is already partially filtered, due to the presence of the charged capacitor across the output terminals. This filtering should be quite adequate for most uses for which you might consider this circuit, but there is no reason why you could not provide more filtering or even some sort of regulating circuitry. It should be mentioned that the output voltage of a voltage doubler circuit is quite load dependent and neither the full-wave nor the half-wave circuit provides much in the way of self-regulation.

There is nothing really critical in the selection of parts for use in a doubler, except that it is definitely recommended to use high-quality capacitors, since they must carry the total load current continuously. There doesn't seem to be any set formulas for calculating the size of these capacitors. In the days of tube usage, it was common to use 40 to 50 μF for C1 and C2, but it has been found in the development of the circuit described here that values of 220 μF have provided adequate current levels. The working voltage of the capacitors should be at least twice the input voltage for C2 plus a safety margin, while C1 can be specified for use at the input voltage. Since the circuit is capacitively loaded, you must also use diodes of good quality because they will see relatively high surges. Never consider using anything less than something in the 1N4001 series with their 1A continuous rating and correspondingly higher surge rating. The resistances shown in Figs. 2-3A and 2-3B are basically there to simulate the source resistances from the transformer and diodes, although you could insert some surge limiting resistance if desired.

Up to this point, a type of circuit which will give a positive output with respect to common line has been discussed. Remembering that many devices require negative voltages, this

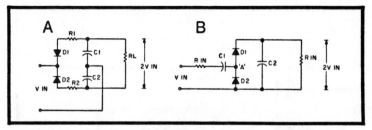

Fig. 2-3. Doubler circuits. (A) Conventional full-wave types. (B) Cascade or half-wave type.

can be accomplished by taking the basic cascade or half-wave doubler circuit and reversing the polarity of everything.

The details of this procedure are shown in Fig. 2-4. This is the schematic of a power supply that has been used with considerable success in a number of projects which have been custom made. The upper 5V section is quite conventional, with an NPN series-pass transistor driven by another NPN transistor set up as a "differential amp." Much has been written regarding this type of circuit, so it will not be covered in detail, except to say that this particular one uses a 2N3055 for the series-pass transistor, and the load draws a continuous 3A from it without any problems; however, hefty heat sink is used for the 2N3055. Proper selection of the transformer, diodes, series-pass transistor, and heat sink would allow you to draw almost any current level you might desire.

The lower section is the one to be more concerned with. If you will compare it to Fig. 2-3B, you will see the similarity. It is the mirror image of the half-wave doubling circuit and, as a result, provides a negative output. You can connect the input to the doubler to either side of the transformer secondary, but just be sure you connect it in front of the rectifiers for the other section of the power supply. The actual voltage on the more negative end of C2 will depend on the secondary voltage of the transformer you choose to use, but you can essentially assume that the unloaded voltage at this point will be $2V_{IN}$ peak (i.e., rated secondary voltage × 1.414). Two zener regulators have been added to this negative voltage to obtain the two levels necessary to supply the MOS/LSI chip. At very low current levels (10—15 mA or less), there is no need for a filter

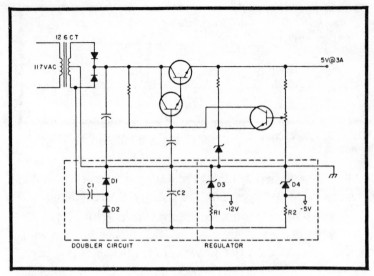

Fig. 2-4. Schematic of a 5V 3A supply with negative 5V and 12V sources from a voltage doubler.

capacitor on these regulated voltages; however, if you wish to draw more current, they would most likely be necessary. If you understood the previous explanation of the mechanics of a voltage doubler circuit, you should be able to see the workings of this circuit with very little study.

To expand on the possibilities of this type of circuit, look at Fig. 2-5. Here you have a power supply which provides 5V at high current levels, −5V and −12V for the MOS/LSI devices, and also plus and −12V to power an operational amplifer. This circuit is very well suited for use with low-power operational amplifiers because, even though the output has rather high ripple levels, most low-power op amps have a very high power supply ripple rejection capability. All this from a transformer with a single 12.6V secondary.

Should you require higher voltages, you could easily employ higher orders of voltage multiplication as long as they have a common input and output. However, it is recommended that you spend a little time in the books before you attempt these levels so that you understand what you can and can't do. Many libraries have a small section on electronics, and you might be fortunate enough to find something there.

You could build this supply with almost any type of construction practice with which you are comfortable. Start by laying out a small printed circuit board to fit what space you have available which holds all parts except the transformer and the series-pass transistor; the transistor is mounted on its own heat sink or, in some cases, sinked to the chassis. In supplies providing lower power levels, it is suggested that you mount the power transistor right on the board in a small heat sink. If you do this, you must watch that the temperature of the device will not rise high enough to damage the board itself or cause problems with some of the other parts on the PC board.

VERSATILE POWER SUPPLY

The circuit is what is termed a versatile power supply. Versatile because all of its many uses—a 12V version powers mobile equipment, a 5V version powers logic circuits, and a 15V version runs operational amplifiers. All in all, there has been nine versions of this circuit tried and tested.

Whether you need a supply to charge batteries, run a portable tape recorder or radio, operate relays, run your mobile FM rig in the house, or power op amps or logic, this

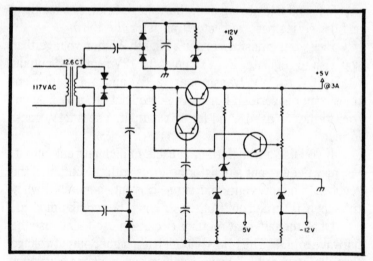

Fig. 2-5. Schematic diagram of basically the same supply of Fig. 2-4 with the addition of a positive 12V line.

123

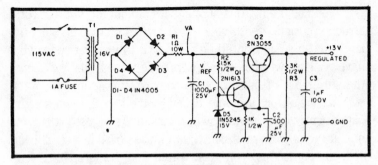

Fig. 2-6. Schematic of 13V 2A supply using NPN transistors.

circuit definitely deserves a try. It's simple, uses a minimum of easily obtainable parts and, best yet, it works.

The basic circuit (Fig. 2-6) is a simple two transistor regulated supply. One transistor, Q1, acts as a reference voltage source and the other, Q2, acts as a series pass regulating element.

The circuit is the same no matter what the output voltage. Only the transformer, zener diode, resistor R2, and possibly Q1 and Q2 have different values.

Circuit Theory

The circuit consists of three sections—the transformer, rectifier, and filter being one, the voltage reference another, and the series pass regulator transistor the third.

Pick your transformer for a slightly higher voltage than you wish to regulate. For example: for 12V regulated output, use a 16V to 19V transformer (a 6V and a 12V filament transformer in series). For 5V regulated output, a 6V filament transformer is used. And for 15V output, use a 24V transformer.

If you use a transformer capable of high current output, you may need to put a resistor in series with the output of the rectifier. This prevents the surge current, generated when the supply is turned on, from destroying the rectifier diodes.

Of all the different rectifier circuits in use today, use the full wave bridge. The bridge circuit has a higher output voltage than the standard full wave rectifier and a higher frequency ripple than the half wave, making it easier to filter; however, any type rectifier, as well as the conventional voltage doublers and triplers, can be used with no circuit degradation.

For output filtering, use a 1000 μF capacitor for C1. This provides adequate filtering. The more capacitance, the more filtering, so you can increase this value if you wish. The regulator also acts as a capacitance multiplier so that the total capacitance is the capacitance of C1 plus the capacitance of C2 multiplied by the gain, h_{fe}, of Q2:

$$C_T = C_2 (h_{fe}) + C_1$$

Resistor R2, the zener diode, and Q1 form a voltage reference circuit. Set the reference voltage to about 1.5V above the desired output voltage. The reference voltage is higher because there are two diode voltage drops between this point and the output. To determine the value of the zener current limiting resistor, R2, you must first know the gain of transistors Q1 and Q2.

The DC gain of Q1 and Q2 can be found in a transistor specification manual. Gain, sometimes called beta or h_{fe}, is the ratio of the collector current to the base current that caused it. For example, a transistor with a gain of twenty, 1 mA of base current causes 20 mA of collector current.

Now, knowing the transistor gains and the desired output current, you pick a value for R2 by calculating how much current must flow from the base of Q1 to make a corresponding amount of current flow from the base of Q2, to cause the desired output current flow into Q2. Simple. Right?

For example, we wish to have a regulated output voltage of 13V at 2.0A. The gain of Q1 is seventy and the gain of Q2 is twenty. The voltage at point A is 20V. To cause 2.0A of current to flow into Q2, the current flow from the base of Q2 must be 100 mA.

$$I_b = \frac{I_{out}}{gain} = \frac{2.0A}{20} = 100 \text{ mA}$$

And similarly, to cause 100 mA of current flow into Q1, the base current must be

$$I_b = \frac{100 \text{ mA}}{70} = 1.4 \text{ mA}$$

If you use a 15V zener diode, the voltage drop across R2 is 5V, so from Ohm's law:

$$R = \frac{E}{I} = 5V/1.4 \text{ mA} = 3.7K$$

Let R2 be 3.3K. The reason for this lower resistance is that most transistors have less gain than that listed in the specification book, so by using a smaller resistor more current will be available, making up for possible low transistor gain. This also gives you a safety margin, in case you need just a bit more current than you thought.

In picking transistors for Q1 and Q2, not only must you pick a transistor with suitable gain but you must also choose it for its type (NPN or PNP), emitter-collector breakdown voltage BV_{ceo}), collector current (I_c), and power dissipation.

All of these circuits use NPN transistors. If you wish to use PNPs, reverse the polarity of the rectifier output, the filter capacitors and the zener diode. Isolate everything from ground. If you require that the negative lead be grounded, ground the emitter of Q2 (see Fig. 2-7).

BV_{ceo} is the voltage at which the collector to emitter junction breaks down. For Q1 this voltage rating must be high enough to stand the difference in voltage between the rectifier output and ground, and for Q2 the rectifier output and the regulated output.

The collector current (I_c) rating of each transistor is the maximum continuous collector current that the collector to emitter junction can safely pass. For Q2, this is the total output current which you require from the supply.

Power dissipation is the maximum amount of power that the transistor can dissipate before it is destroyed. The power dissipation rating is usually given for an ambient case temperature of +25° C. If you heat sink the transistor (which is recommended for Q2), you can exceed this dissipation by an amount which depends upon how well your heat sink dissipates the power.

The output circuit consists of R3, a bleeder resistor chosen to allow a couple of milliamperes of current to flow, and

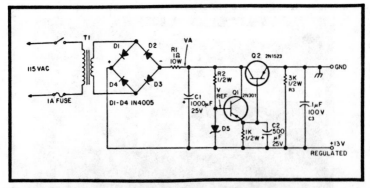

Fig. 2-7. Schematic of 13V 2A supply using PNP transistors.

capacitor C3, which acts as a high frequency filter to keep any zener noise or voltage spikes out of solid-state equipment.

Construction

The circuit layout is not critical and almost any type of configuration can be used. The supply has been constructed both on a piece of pegboard with the device leads serving as hookup wire, and on a printed circuit board. With the circuit board shown in Fig. 2-8, the electrolytic capacitors are mounted external to the board due to their large physical size. This figure also shows Q2 mounted on the board. While this will work for low power applications, it is suggested that you heat sink this transistor, as it does pass the total load current and can get warm.

You will notice that there is a break in the land coming from the positive output of the bridge. This is where R1, which must be heat sinked externally from the board, is connected. If you do not use R1, connect a jumper in its place.

A resist marking pen was used to draw the circuit on a PC board. You can do the same or make a photographic negative and use the photoresist method of making a board. The board in Fig. 2-8 is shown full size. The only precaution is to mount the heat sinked transistor, Q2, where its case cannot be accidentally shorted to ground. If possible, mounting it on an attachable heat sink and mounting the heat sink inside the box (with a few ventilation holes) will work fine.

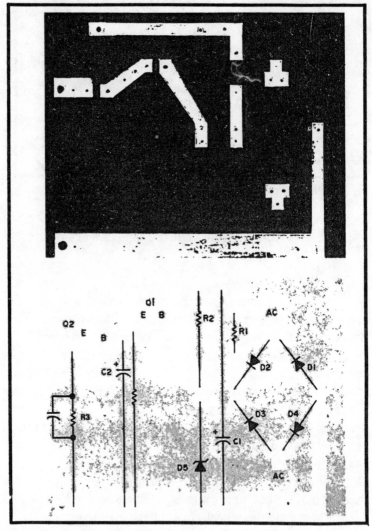

Fig. 2-8. PC board layout for power supplies shown in Figs. 2-6 and 2-7.

SIMPLE DUAL-VOLTAGE SUPPLY

This regulated dual voltage power supply (Fig. 2-9) can start you off experimenting with the more common TTL integrated circuits and the increasingly popular CMOS integrated circuits. In addition, the higher voltage available from the power supply is good for a vast number of linear IC projects.

The power supply uses two type 309K three-terminal regulator integrated circuits. These consist of an in, out, and common pin connection, thus this IC is the simplest possible device to work with. The K-suffix designates the type TO-3 package. This regulator is also available in the type TO-5 package, but we will use the higher power package to obtain output currents in excess of 1A. This is a particularly fine power regulator specifically built for 5V output for TTL use, but a simple connection allows the regulator to furnish higher output voltages with equally excellent regulation.

The 309K three-terminal regulator is rated by the manufacturer at output currents in excess of 1A and employs internal current limiting, thermal shutdown, and safe-area compensation. All of this means that it is essentially indestructible. It also does not require a lot of external components, unlike most other regulator circuits. It requires only two resistors to provide a higher output voltage. All told, this is one of the easiest to build fully regulated and self-protected power supplies around.

There are no critical layout precautions. If the supply should exhibit a tendency to oscillate as evidenced by erratic operation, simply connect a 0.22 μF capacitor directly from input pin one to ground as close as possible to the regulator.

If you can scrounge up some of the parts from your junk box, you are already ahead of the game, but later on you will see how all of the parts can be obtained from a source near you to make it as easy as possible for you to get started experimenting. When all parts are on hand, you can assemble and package them in any way you like.

The power supply schematic (Fig. 2-9) may look a little strange to you, but do not be alarmed. This is a solid-state version of the so-called economy power supply, more popular several years ago than it is today. It is a combination of the well known bridge and full wave rectifier circuit configurations. One regulator is connected to the transformer secondary winding center tap to yield 5V at the output for TTL circuits, and makes use of two of the diodes in the bridge rectifier to work as a standard full wave rectifier circuit. The other regulator is connected to the output of the bridge rectifier to yield about 14V at the output for CMOS and linear circuits. This regulator is biased up from ground by the two resistor divider network across the output. The regulated output voltage is adjustable by virtue of the variable resistor. The regulator requires a minimum voltage differential of about 2V between the regulated output and the DC voltage at input pin 1 from the output of the rectifier diodes and filter capacitor.

Most CMOS integrated circuits are rated by the manufacturer at 15V maximum, and it is wise to limit the output voltage of the power supply to not over, say, 14V. A voltage between 12V and 14V is good in order to take full advantage of the exceptionally high noise immunity of CMOS integrated circuits, and is also a good operating voltage for linear IC projects.

The variable resistor may be replaced with a fixed resistor of equal in-circuit value, if it is not desired to vary the output voltage. Temporarily connect a variable resistor and adjust it to set the regulator output voltage as read by a voltmeter. Shut off the power supply, remove and measure the value of the resistor, and solder in a fixed ½W resistor of the nearest standard value.

Notice that the common terminal of the 309K is also the case of the regulator, so it is necessary to insulate the higher voltage regulator from the heat sink. For this purpose use the insulator found in the power transistor mounting hardware kit. Spread a very thin coat of heat sink compound (silicone grease) on each side of the insulator before mounting the regulator to the heat sink. Use the power transistor sockets to save soldering directly onto the pins of the regulator. This usually results in a messy-looking job. Perhaps a more important reason for using the sockets is that they have self-aligning insulated hubs that center the pins as they pass through the holes in the heat sink. The socket also insulates the mounting screws from the heat sink and prevent shorts from this cause.

Building this power supply is about as simple as can be. All of the parts are available from Radio Shack. Any equally rated part can be substituted to build the power supply.

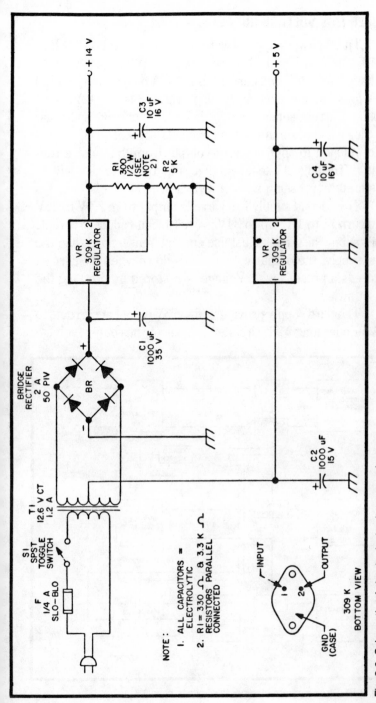

Fig. 2-9. Schematic of simple dual-voltage supply.

131

SUPER LOW-VOLTAGE SUPPLY

Three power supplies were constructed as shown in Fig. 2-10. The first is a dual tracking supply with variable output voltage 0 to ±20V and current to 100 mA on each output (200 mA total current capacity). Also available is a +12V, −6V option. Current sensing is done in both the positive and negative legs, and when the current exceeds a preset level, a signal is developed to shut down the output from the voltage regulator. This signal latches so that output voltage can only be restored by pressing a reset switch.

The second supply has variable output from 2.6V to 25V and current to 1A. Up to 34V is available at reduced current. This supply also has adjustable current sensing and, like the first supply, the output voltage shuts down when the current exceeds a preset level. Voltage is restored by pressing the reset switch.

The third supply provides a fixed 5V output at currents to 1A for operating TTL circuits. This supply has output voltage

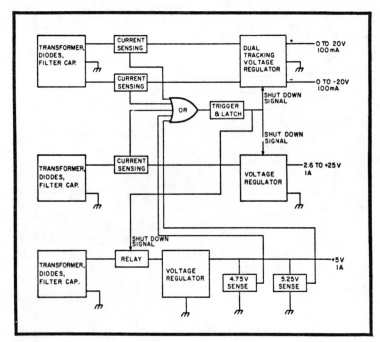

Fig. 2-10. Block diagram of super low-voltage supply.

sensing and will shut down if the voltage moves outside a preset range from 4.75V to 5.25V.

The first supply provides the power for the sensing circuits used in all three supplies. Also, if any one supply shuts down, the other two will shut down also.

All three supplies use voltage regulators that are short-circuitproof, an added safety bonus in the event that the current sensing circuits are manually disabled or in the event of the failure of some component in the current sensing networks.

Current Sensing

The current sensing network in Fig. 2-11 operates as follows: Assume that initially no current is drawn from the supply. With R2 set to 500Ω, R2 + R3 = 21K and R4 + R5 = 21K. With the wiper of R4 set closest to R3, the voltage at pin 11 of voltage comparator IC1A will be 14V, exactly half the voltage across C1. Assuming for the moment that no current flows in R1, the voltage across R6 and R7 will be 28V and the voltage at pin 10 of IC1A will be 14V also. When current is drawn from the positive leg of the supply, a voltage drop develops across R1 and the voltage at pin 10 of IC1A drops below 14V. This drives pin 13 of IC1A positive and the resulting current in R21 charges C3. Q1 fires, sending a pulse through C4 to SCR1. SCR1 turns on, operating relay K1 and forcing Q2 to switch on. Q2 shorts out R27, thus reducing the output of IC3 to nearly zero. K1 interrupts the current to IC6 in Fig. 2-12. Q1 also sends a pulse to C14 in Fig. 3. This pulse turns on SCR2, forcing Q4 to switch on; this action reduces the output of IC5 to zero.

When the load is removed from the output of IC3, the power can be restored by opening S2A and S2B (normally closed switches). By moving the wiper of R4 closer to R5, the voltage at pin 11 of IC1A is lowered. It then requires a greater voltage drop across R1 (more current in the load at output of IC3) to lower the voltage at pin 10 of IC1A so that pin 13 will go positive. Thus the setting of the wiper of R4 determines what current will drive pin 13 of IC1A high.

An identical network consisting of R8 to R14 and IC2 senses the current in the negative leg of the supply. The output of IC2 switches between zero and $-26V$ approximately. Since IC1B will not operate normally with any input below $-0.3V$, the voltage from pin 6 of IC2 is divided down by R15 and R17 so that the voltage across R17 switches between zero and $-0.25V$.

R16 and R18 form another voltage divider which provides $-0.15V$ to pin 8 of IC1B. Thus IC1B switches like IC1A in response to an overcurrent in R14. D5 and D6 form an OR gate, hence isolating the outputs of IC1A and IC1B from one another.

In Fig. 2-12, current sensing is done in the same manner as described for the positive leg of Fig. 2-11. Since the maximum current for this supply is ten times greater than for the first supply, resistance values have been adjusted accordingly. D9 forms another part of the OR gate that feeds R21.

Voltage Sensing

For the 5V supply in Fig. 2-13, it is more desirable to have output voltage sensing than current sensing. This is because there are wide variations in the current demanded by TTL circuits when they are switching from state to state. The current limit point would always have to be set rather high, and consequently only gross overcurrents could be sensed. On the other hand, a circut that senses when the voltage falls below 4.75V, the lower operating limit for 7400 series TTL, is quite useful.

Suppose, for example, that you are operating near the 1A limit of IC6; a brief current pulse could exceed this limit and the internal circuit of IC6 would then allow the output voltage to drop. Without voltage sensing this could easily go unnoticed and your circuit would malfunction.

In Fig. 2-13, D14 provides a reference voltage. R41 acts as a voltage divider and is set to 5.25V. R42 is another voltage divider and is set to 4.75V. IC1C and IC1D compare the output of IC6 to these voltages and, if the output moves outside the window from 4.75V to 5.25V, pin 1 or pin 2 will go

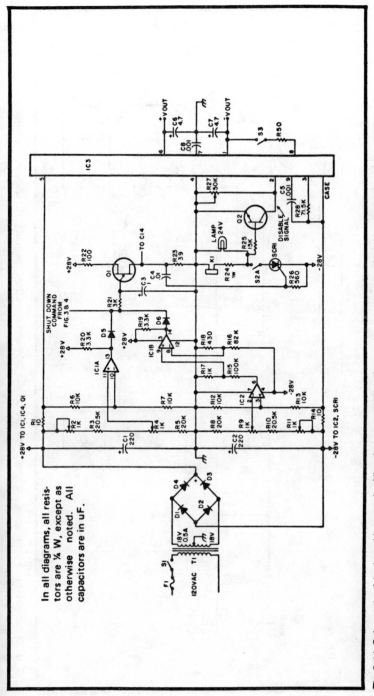

In all diagrams, all resistors are ¼ W, except as otherwise noted. All capacitors are in uF.

Fig. 2-11. Schematic of dual-tracking supply. All resistors are ¼ W and capacitors are in microfarads unless otherwise noted.

135

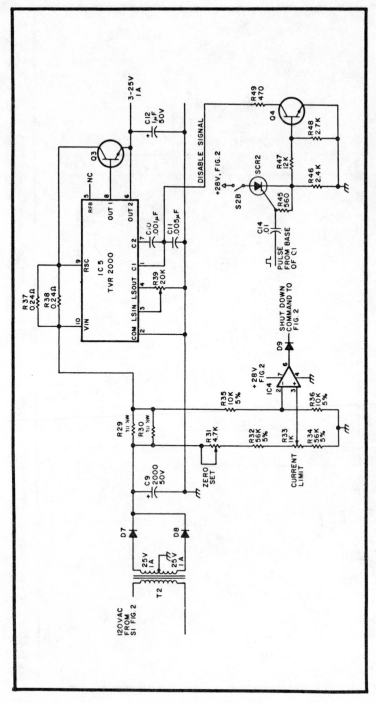

Fig. 2-12. Schematic of variable voltage supply. All resistors are ¼W and capacitors are in microfarads unless otherwise noted.

high. This signal goes to R21 of Fig. 2-11 and eventually shuts down all the supplies.

Response Time

R21 and C3 determine the response time of the circuit. With R21 equal to 3K and C3 equal to 1 μF, the circuit responds to an overcurrent, overvoltage, or undervoltage that lasts 3 milliseconds (ms) or more. K1 adds an additional 7.5 ms to the time required for the 5V supply to shut down. By reducing C3 to 0.1 μF, response time can be made as low as 0.3 ms. R21 can be increased to as much as 10M if desired to lengthen the response time, but should not be reduced below 3K.

The Voltage Regulators

The 4194TK regulator is available through many sources. It is internally current limited at about 350 mA when the positive output is shorted to ground. It also has internal thermal limiting that will reduce the output when it gets too hot. A small heat sink is required when the operating current is 100 mA in each leg of the output.

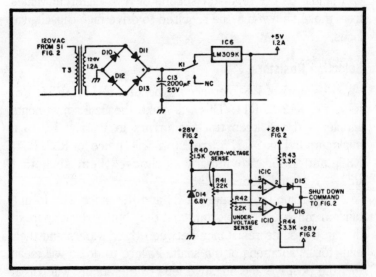

Fig. 2-13. Schematic of 5V supply. All resistors are ¼W and capacitors are in microfarads unless otherwise noted.

In Fig. 2-11, S3 is normally open. When S3 is closed, R27 can be adjusted to give +12V, −6V output for the operation of certain types of voltage comparators.

The 309K also has current limiting and thermal limiting. It will provide a little over 1A when mounted on a heat sink with the circuit shown.

The TVR2000 has been available from Poly Paks for a number of years and is quite inexpensive. It is surprising that in spite of its outstanding performance and low cost never once has it been written up in a monthly publication. Perhaps it is because the information on how to use it is hard to find; the specification sheets that come with it do not give enough information on how to use it.

In Fig. 2-12, the foldback current limiting option is not used. Instead, simple short-circuit sensing is used. R37 and R38 set the short-circuit current to a value of about 1.2A. The relationship here is

$$R_{sc}I_{OUT} \approx 0.1V$$

where R37 and R38 in parallel make up R_{sc}. R39 sets the output voltage. Q3 acts as a current booster and is mounted on a heat sink. C10 stabilizes the current limiting circuitry and C11 stabilizes the regulator section of IC5. Different values from those shown may be required to drive high capacitance loads.

Selecting Resistors

Resistors of 1% tolerance are best for R1, R3, R5 to R8, R10, and R12 to R14. This will make the final adjustments simpler and will keep tracking errors in R4 and R9 to a minimum. In Fig. 2-12, 5% resistors will suffice for R32, R34, R35, and R36, providing you choose them such that R32 ≤ R34 and R36 ≥ R25.

Regarding the tracking of R4 and R9: since they form a tandem control, it is important that they both exhibit approximately the same resistance between their wipers and their ends for all rotations of the shaft. Failure to do so will mean that the positive and negative legs of the supply will trip at different currents. Several dual controls bought did not track

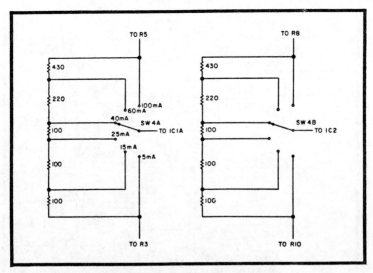

Fig. 2-14. Switch selected resistors replaced by R4 and R9.

very well. If you want very good tracking, replace both R4 and R9 with a series of 5% resistors and use a two pole rotary switch to select the current limit you want as shown in Fig. 2-14.

Construction

All three supplies were constructed on a single 4- by 5-inch printed circuit board as shown in Figs. 2-15 and 2-16. IC3 does not plug directly into the board; the holes in the board have been spaced out to assure clean etching. Solder a short wire to the outside of each pin of IC3; insert the wires into the PC board and solder. A piece of aluminum was bolted to IC3 as a heat sink.

There are so many connections to the PC board from the external switches, controls, transformers, etc., that it was not possible to arrange for an edge connector on a board of this size. Instead there are about 35 wires soldered at various points around the edge of the board, and all are routed to one end of the board so that the board can be hinged outward from the chassis if parts on it need to be replaced in the future.

All components fit nicely on a chassis 10- by 6- by 2-inches.

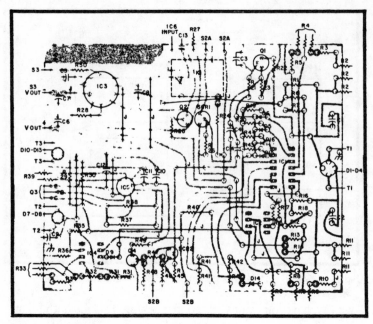

Fig. 2-15. Parts layout.

Final Adjustments

Switch S2 to reset. Leave S3 open. This disables the shutdown mechanism. Connect a high impedance voltmeter between pin 7 of IC1D and ground. Adjust R42 for a reading of 4.75V. Connect the voltmeter between pin 4 of IC1C and ground. Adjust R41 for a reading of 5.25V.

Set the wiper of R33 to the end closest to R32. Connect a voltmeter between pin 6 of IC4 and ground. Adjust R31 so that the reading just goes to zero.

If you are using a dual potentiometer for R4 and R9, proceed as follows: Set the wiper of R4 to the end closest to R3; the wiper of R9 should then be at the end closest to R10. Connect a voltmeter between pin 13 of IC1A and ground. Adjust R2 until the reading just drops to zero. If you run out of adjustment with R2, interchange R6 and R7 and try again.

Connect the voltmeter between pin 14 of IC1B and ground. Adjust R11 until the reading just drops to zero.

If you elect to use the switched resistors in Fig. 2-13, proceed as follows: set the switch in Fig. 2-13 to the 5 mA position. Connect a load between the positive and negative output terminals of the supply, and adjust the output voltage so that the load draws 5 mA. With a voltmeter from pin 13 of IC1A to ground, adjust R2 until the voltage just drops to zero.

If you run out of adjustment with R2, interchange R6 and R7 and try again. Connect the voltmeter between pin 14 of IC1B and ground. Adjust R11 until the reading just drops to zero.

A partial parts list is shown in Fig. 2-17.

ZERO TO 28V SUPPLY

As you can see from the schematic in Fig. 2-18, the circuit is simply a variable autotransformer feeding a husky

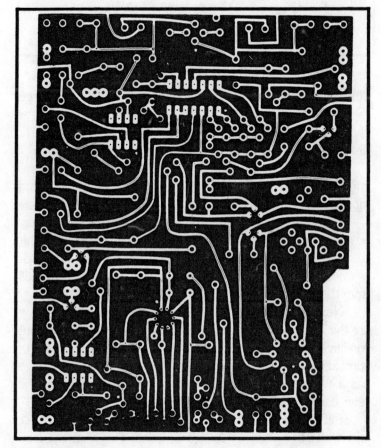

Fig. 2-16. PC board.

Partial Parts List

IC1	339
IC2, IC4	741
IC3	4194TK
IC5	TVR2000 (available at Poly Paks)
IC6	309K
D1-D4 } D10-D13 }	2 A 100 piv bridge rectifier
D7-D8	half of 2 A 100 piv bridge rectifier
D5,6,9,15,16	1N4148
D14	1N957B 6.8 V, 0.4 W zener diode
Q1	2N2646
Q2	2N4249
Q3	MJE3055
Q4	2N5550
K1	ITT type 24A02C18A
R4,R9	dual section control; see text
SCR1,SCR2	C103B

Fig. 2-17. Partial parts list for the super low-voltage supply.

30V transformer into a bridge rectifier. The filter is a large capacitor. It would be nice to have a 30A choke, but the filter shown has been satisfactory for all of the uses previously discussed.

You may skip this part if you have a transformer or are going to purchase one. The transformer for this unit was a modified TV transformer.

First remove the outer shell and the laminations. It may be necessary to use a little force to accomplish this. Never hit the laminations directly with a hammer; use a mallet or a block of wood with your hammer. Take care to preserve the leads from the windings.

Next, remove the secondary windings. Generally they are the outer windings. When you remove the 6.3V filament winding, count the number of turns. It probably will consist of approximately twelve turns. Usually this type of transformer has a two turns per volt ratio. Retain this information for rewinding the secondary.

Finally, rewind the secondary; use AWG #12 copper wire with a thick thermoplastic insulation. This wire was used because there happened to be plenty on hand. Obviously, an

The supply is simple to build in any size of metal enclosure suitable for the components used. The only precautions to observe are to firmly heat sink the LM317 to one side of the metal enclosure and to keep the 0.1 μF capacitor going from pin 3 to ground, the 10 μF capacitor going from pin 1 to ground, and the 120Ω resistor going between pins 2 and 1, *all* connected directly at the LM317 terminals. The other components may be mounted wherever it is convenient to do so.

The zener diode/resistor/LED combination at the output of the supply serves as a crude but useful voltage output indicator without having to build a regular voltmeter in the supply. The LED just starts to glow when the output voltage is about 9V to 10V, depending on the tolerances of the components used. The 1K resistor is adjusted so the LED just glows fully when the *maximum* output voltage is reached. So by using the fixed output voltage positions and watching the LED, one can obtain a fairly good estimate of what the variable output voltage is set for.

NOISE-FREE SUPPLY

The circuit shown in Fig. 2-22 is a basic DC-to-DC converter, but instead of the customary square wave output this

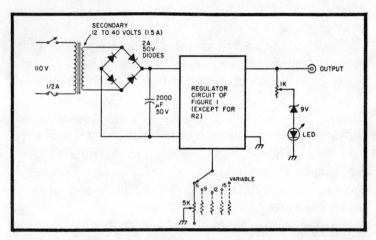

Fig. 2-21. Complete schematic of 1.2V to 37 V supply. The switch simply selects different 5K pots, which are set for 6V, 9V, 12V, and 15V, and an adjustable output. The latter 5K pot is mounted on the front panel. The function of the LED is explained in the text.

147

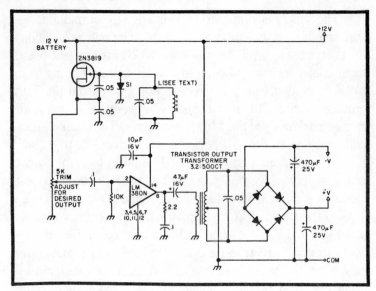

Fig. 2-22. Schematic of noise-free supply.

one starts off with a sine wave. This does away with any possibilities of generating any spikes. The circuit uses an audio amplifier to build up the signal, then feeds it to the power transformer.

The idea worked quite nicely. Typically 5 mA can be drawn from each output, but as much as 10 mA would not cause problems.

All kinds of chokes and transformers were tried in the oscillator tank circuit, but eventually it was found that a hand wound pot core inductor worked best. The pot cores used were obsolete Ferroxcube parts, but similar units should work as well. The ones used here are about ⅜ inch in diameter and 5/16 inch thick with both halves assembled. Material is Ferroxcube 3C. The bobbin was wound with 800 turns of AWG #44 magnet wire. On a homemade bridge, the inductance checks out around 700 millihenries (mH). In the power supply the oscillator frequency is around 900 Hz.

An LM380N audio amplifier IC is used to drive the voice coil side of a standard 500Ω to 3.2Ω output transformer (Radio Shack #273-1379). The bridge rectifier is one of the small plastic units about the size of a TO-5 transistor case. The

148

transformer center tap is grounded, and the dual polarity voltages taken from either side of the bridge. Output level is set by the 5K vertical trimmer, which controls drive to the LM380. This control should be set with the load connected. All decimal value capacitors are 50V discs and the rest are electrolytics. Resistors are ¼W carbon. The silicon diode may be a 1N914 or any other type used for switching or general purposes. Other JFETs will work in most cases as the oscillator transistor. Just make sure you get the right pins in the right holes since not all packages have the same pinouts.

Tests made with ±15V out and 10 mA load on each supply showed a maximum ripple of 15 mV peak to peak. At 5 mA loads, the ripple dropped to 8 mV peak to peak. Input current from the 12V source was 85 mA and 55 mA respectively. This is not particularly good efficiency, but at these low levels it was of no great consequence. The 900 Hz hum was just about audible with the receiver quiet but normally was lost under background noise.

The PC board layout and parts placement are shown in Fig. 2-23. The board is 2.3 inches square. The pot core inductor is potted in a cylindrical form after winding, with two radial leads for insertion into the PC board.

INEXPENSIVE 12V 3A SUPPLY

The circuit (Fig. 2-24) is very straight-forward and simple. It is by no means original, but is a combination taken from several circuits. The great thing about this power supply is that all the parts except the two 10W zener diodes can be obtained from a local Radio Shack store. The entire supply should cost no more than $24.00 if all the parts are purchased new, and a less expensive cabinet or modest junk box will bring this down considerably.

Three of these supplies have been built over the past couple of months and all have worked like a charm. Meters were added to monitor voltage and current so the supply could be used for future projects. Every part was purchased at Radio Shack, except the 10W zeners. These were bought from a local TV parts store, so they should present no problem for anyone wishing to duplicate this supply.

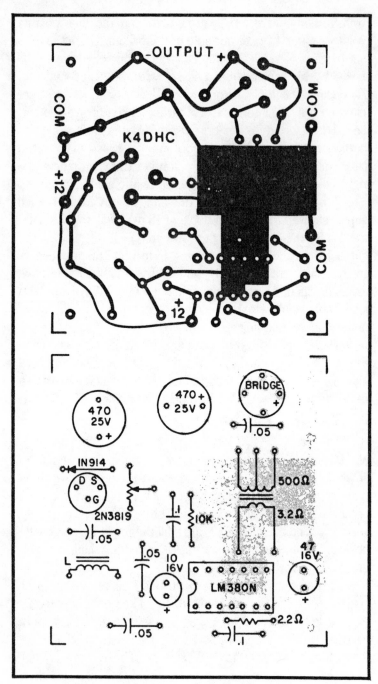

Fig. 2-23. PC board layout and parts placement for the noise-free supply.

150

Parts layout is not critical. The only suggestion is to use a bus line for the negative voltage instead of a common chassis ground. This way a stray screwdriver from the case of Q1 or Q2 to the chassis will not short the transistors. The 40W power transistors come four to a package. A bit of experimenting was required to find a pair that match, but at 30¢ each this is a minor problem. The rear chassis panel works fine as a heat sink; however, for added cooling small individual heat sinks were added.

All in all, this little supply works great, looks good, and is easy and economical to build. The regulation in the 15W mode is less than 0.5V at 2.8A.

FIVE ULTRASIMPLE SUPPLIES

Some time ago there became available to the electronics experimenter a number of three-terminal monolithic integrated circuit voltage regulators. With a minimum of external parts, these ICs can be easily and inexpensively used in the design and construction of high quality, low-voltage regulated power supplies.

In this section the basic use of the LM309 and μA7800 series of regulators are explained. They are all fixed voltage units, most exhibiting a continuous duty current capacity of 1A. Available through many surplus dealers, they cost under $2, and the price is still going down, *in single unit quantities*.

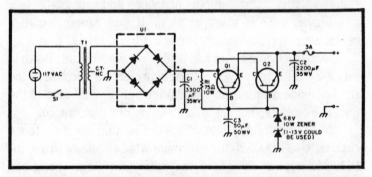

Fig. 2-24. Schematic of inexpensive 12V 3A supply. Q1 and Q2 are 40W power transistors (Radio Shack 276-592); T1 is a 12.6V 3A transformer (Radio Shack 273-1510); U1 is a full-wave 100V 6A bridge rectifier (Radio Shack 276-1171); cabinet is a 5¼- by 3- by 5⅞-inch box (Radio Shack 270-253).

Device	Current	Voltage	Package
LM309H	200 mA	+5 volts	T0-5
LM309K	1.0 Amp	+5 volts	T0-3
uA7805	1.0 Amp	+5 volts	T0-220
uA7812	1.0 Amp	+12 volts	T0-220
uA7815	1.0 Amp	+15 volts	T0-220
uA7818	1.0 Amp	+18 volts	T0-220
uA7824	1.0 Amp	+24 volts	T0-220

Fig. 2-25. Specs of the LM309 and μA7800 series three-terminal regulators.

There are many other regulators available on the surplus market, but it has been determined to describe only the LM309 and μA7800; however, if you spend a little time studying the published data sheets on other units, you will find that most of the circuit design procedures outlined here are directly transferable to use with other three terminal regulators.

Parts Selection

The specifications for the regulators described here are shown in Fig. 2-25.

Figure 2-26 shows a practical power supply circuit, using any one of the regulators in Fig. 2-25. It needs no additional parts. It will work with all of these three-terminal devices by simply varying a few component values and ratings.

Figure 2-27 is also practical: If you cannot locate a center-tapped transformer, replace T1 and D1-D2 from Fig. 2-26 with the transformer and bridge rectifier circuit shown in Fig. 2-27. The full-wave rectifier is preferred, because of the smaller diode drop of the two diodes in Fig. 2-26 than of the four in Fig. 2-27, but the difference is not that crucial.

Now let's assign some values to the components in those circuits. Use Fig. 2-25 to determine what regulator to use for IC1. Make your choice on the basis of voltage rating and current capacity.

Refer to the graphs in Fig. 2-28. There are five graphs for use in determining the T1 transformer voltage and C1 filter

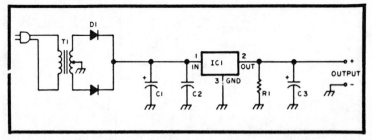

Fig. 2-26. Full-wave version of ultrasimple regulator with a center-tapped transformer. Set R1 5mA for current. That is, regulator voltage divided by R1 equals about 5 mA. Examples are 5V/1K, 12V/2.2K, 15V/2.7K, 18V/3.3K, and 24V/4.7K.

capacitance for five separate 1A regulators. The transformer voltage is for a center-tapped transformer with a full-wave rectifier, as in Fig. 2-26. If you use the bridge rectifier configuration from Fig. 2-27, halve the transformer voltage but remember to *add* 1V to compensate for the voltage drop of the extra diodes. If you use the schematic from Fig. 2-26, the current capacity of T1 need only be 0.5A, or half of what you want the completed supply to deliver. The lower voltage transformer needed for the less efficient bridge of Fig. 2-27, however, must be rated the full 1A.

Filter capacitor C1 should be an aluminum computer-grade electrolytic. If you have any question concerning interpolating a capacitance to use with any given transformer, round *up*. This will help to make up for component tolerances and line voltage fluctuations. Since readings off the graph are minimum allowable capacitances, you might want to use a larger value capacitor anyway.

Since the graphs were calculated using 1A of drain, if some fraction of 1A is to be drawn, the C1 value can be

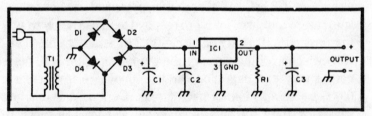

Fig. 2-27. Bridge version of ultrasimple regulator without using a center-tapped transformer. Keep in mind the extra forward voltage drop of the diodes.

153

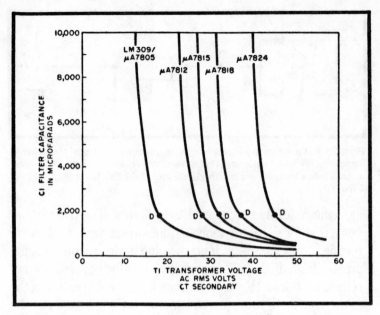

Fig. 2-28. Graphs for determining the value of C1 and the voltage of T1. Point D is the dissipation limit.

reduced by that fraction. For example, if only 100 mA is needed, then only one-tenth the capacitance is necessary, or any other such fraction. It works the other way around, also; if you make a 10A supply (with a pass transistor), you will need ten times the filter size.

Exercise prudence when calculating capacitor voltage ratings. Use at least one and a half times, but preferably twice, the RMS transformer voltage across the capacitor.

Do not omit C2. It is an RF bypass capacitor, especially necessary if the power supply is to be used in an area of strong RF, such as a ham shack, but still needed even if used elsewhere. C2 helps to absorb the spikes and glitches on the line that are too sharp for C1 to respond to. It can be anywhere from around 0.10 μF to 0.33 μF—0.22 μF is a happy medium.

Selection of the value of the output capacitor (C3) need not be very exacting. You can use anywhere from around 0.005 μF to a couple of microfarads. Generally, use larger values for larger currents. A common value for a 1A supply is about a 1 or 2 μF tantalum. Some of the the manufacturers'

154

data sheets insist that it is not needed, but, since the very next phrase in the paragraph is always that its presence helps to improve transient response, it is suggested not to omit it.

These regulators do not work well at low current drains of 5 mA and under. Therefore, R1 has been calculated to draw about 5 mA from the supply. A half- or quarter-watter is sufficient. A pilot light (LED or incandescent) could be wired in to serve the same purpose. If the supply is being built into a project so that the drain is never allowed to go below 5 mA, R1 could be altogether eliminated.

The choice of rectifiers for D1 through D4 is not at all critical. The graphs were compiled using 1.3V as the forward diode drop. This figure is based on the 1N4000-series silicon 1A rectifiers. It should be close enough for interchanging with any of their surplus equivalents or replacements. Types 1N4002 or 1N4003 diodes would be a good choice for any of the transformer voltages on the graphs. Some of the lower voltage units can even get by with 1N4001s. Just make sure that you don't exceed the diodes' PIV ratings.

Automotive Applications

There are many times when you want to operate a transistorized device in your car that requires a supply voltage different from that of the standard 12V automotive electrical system. A higher voltage requirement would take a power convertor exceeding the scope of this section, or a separate battery. But these IC regulators are ideal for obtaining lower voltages from the 12V electrical system. Figure 2-29 is a supply circuit tailored specifically for use in a car.

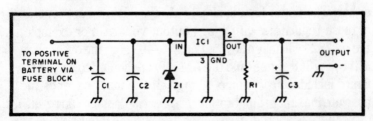

Fig. 2-29. Automobile version of the ultrasimple regulator where transients are apt to be numerous. Z1 can be anywhere from 24V to 30V, whatever is in your junkbox.

155

An automotive environment is, however, very prone to transients. Special precautions should be exercised to protect the regulator. Positive spikes on the input line can very easily cause that input terminal to go above the maximum allowable input voltage for the IC. Therefore, the zener diode (Z1) has been put into the circuit to help shunt any higher spikes to ground. You can use whatever zener you have in your junk box in the 24V to 30V range.

The rectifiers are absent in this version, but note that the filter capacitor is still there. Keep it there; your car's alternator puts out an awful DC waveform. Filtering is still needed.

All of the other parts are the same as the other circuit diagrams.

Construction Considerations

The construction considerations you need concern yourself with when using these IC regulators are not very numerous. But those that do exist should be carefully observed.

There are definite reasons for the endpoints of the graphs in Fig. 2-28. *Do not extrapolate transformer voltages from either end of the graphs.*

These regulators will only work when an input voltage between certain limits is present at the input terminal. The dropout voltage, or level below which the unit will not operate properly, is usually 2V greater than the rated output voltage. Thus the LM309 needs an input of at least 7V to work correctly. If the input is too low, it will not regulate at the desired voltage and the ripple will be excessive.

The input level to these regulators cannot be too high, either. The upper limit is almost always 35V, except in some of those regulators with a high output voltage. For example, the μA7824 will taken an input of up to 40V, but the other units in the μA7800 series (and the LM309) are limited to 35V. This figure includes *peak* ripple voltage. If the input exceeds the absolute maximum input rating, *permanent* damage can result.

There is another reason for keeping the input voltage as low as possible. The dissipation of the devices in the TO-3 or

TO-220 packages must be limited to around 5W to 10W, depending upon heat sink efficiency. (The LM309H, in its TO-5 metal can package, is limited to around 1W or 2W power dissipation.) In this case, exceeding the junction temperature limit, for whatever reason, will only trigger the internal protective features—thermal shutdown and current limiting—causing no permanent damage. This is especially important if you want to draw a full ampere from the supply. Although the dissipation capacity is still a function of heat sink efficiency, 20W is the absolute maximum dissipation regardless of how much heat you remove from the case.

For these reasons point D (dissipation limit) has been included on the graphs of Fig. 2-28. If you want to draw high currents from the IC, use a transformer voltage to the left of (above) points D and attach the regulator to an adequate heat sink with liberally applied silicone grease. The Signetics LM309 data sheet indicates that a typical commercial heat sink, the Wakefield 680-75, should be sufficient to allow the TO-3 package of the LM309K to dissipate from 5W to 7W at room temperature before thermal shutdown is triggered. Use this as a guideline for determining your own heat sink requirements. Keep in mind that placement of the dissipation limits in Fig. 2-28 is in no way absolute. Dissipation capacity is very dependent upon heat sink size and efficiency. Simple experimentation should be used to determine what size is needed for your specific application.

Refer to the base diagrams in Fig. 2-30 for finding the pin connections of the TO-5 metal can, TO-3 metal power, and

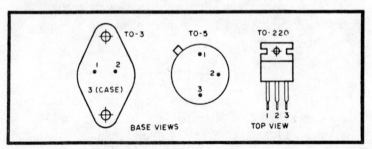

Fig. 2-30. Basing diagrams for the TO-5 metal can, TO-3 metal power, and TO-220 plastic power packages.

the TO-220 plastic power packages. Follow normal wiring procedures, with no other special precautions except to make sure that you use large enough wire for the current you intend to draw.

Design Calculations

For those of you who want to apply these design procedures to other three-terminal regulators, or just simply want to know a little of the theory behind this shortened design method, here are all the necessary calculations and a little on why they are used:

You do not get exactly 12.6V DC from a 12.6V transformer. Nor do you get 6.3V; in fact, there is no such easy one-to-one relationship. Rather, you get a fluctuating DC waveform whose amplitude varies all the way from zero to a peak value of $\sqrt{2}$ (or about 1.414) times the transformer voltage.

As you increase the size of the filter capacitor, the ripple stabilizes. The waveform approaches an inverse sawtooth—an almost vertical charge time, followed by a nearly steady linear discharge rate, all repeating every 8.333 milliseconds (cycling at 120 hertz). The peak value of the waveform remains where it was, but the amount of ripple subtracted from it decreases. Approximating a sawtooth will be close enough for our purposes.

Since specific transformer voltages are a lot easier to come by than are capacitor sizes (especially if you rewind them yourself, but that's another whole story), calculate the necessary transformer size needed with a given capacitor. Let's start by taking that capacitor size and calculating its ripple per given current drain:

$$V_r = \frac{8(I)}{C}$$

where V_r = peak-to-peak ripple voltage, I = current drain in milliamperes, and C = filter capacitance in microfarads.

You now have the amount of ripple, the diode drop, and the dropout voltage of the IC regulator. The sum of these voltages equals the minimum necessary peak transformer value. The *minimum* RMS voltage that you can use therefore is:

$$V_t = (1.414) \ (V_d + V_r + 1.3)$$

where V_t = transformer voltage, volts center-tapped (RMS), V_d = regulator dropout voltage (usually $2 + V_{out}$), and V_r = the ripple voltage calculated above.

Since this equation is for use with a center-tapped transformer and a full-wave rectifier, here is a modified formula to use for Fig. 2-27's bridge. It is not simply half of the full-wave value, so use this modified formula instead of trying to work it out of the other one:

$$V_t = (0.707) \ (V_d + V_r + 2.6)$$

where all variable names are the same as before.

If you are using other rectifiers and have more accurate data on them, you can insert your diodes' forward drop in place of the 1.3 and 2.6 that was used for 1N4000 series rectifiers. Remember that, whenever the *average* current, the peak value is much higher at specific instances. Currents above 10A can be in a 1A power supply. Keep this in mind if you calculate your own rectifiers' forward drop.

Round up your final answer. This will incorporate a safety margin to allow for component tolerances and line voltage fluctuations. A 10% or 20% round up margin could safely be used, providing that you do not approach the absolute maximum input ratings of your device. Your final calculation should be a check to make certain that the instantaneous voltage, with the actual component values you have decided to use, never exceeds the capacitor or regulator input ratings. When you round up, all that you in fact are doing is improving the line regulation at only a slight trade-off in size and efficiency.

Device Availability

The LM309 regulator (both the H- and K-package styles) has been designated to be worst case 7400 TTL compatible. Barring thermal shutdown or current limiting, their output voltage will never deviate outside the 4.75V to 5.25V range that 7400 TTL logic requires, even counting line, load, and thermal regulation summed worst case.

The LM309H is therefore a prime candidate for on-card regulators for logic circuits needing less than 200 mA. And it is in fact somewhat unique in this respect. The larger package styles, although still small enough for PC board mounting, are more conducive to chassis mounting.

The LM309 is by far the most readily available of the many three-terminal voltage regulators. For the sake of printing costs, some suppliers have omitted the LM prefix in their advertisements. Others do not make obvious the distinction between the H- and K-case styles. Nevertheless, a good share of the surplus dealers that advertise in the back of experimenter-oriented electronics magazine do carry the LM309.

A different story exists concerning the μA7800 series. Even though here again some distributors have dropped the prefix in their advertisements for the sake of brevity, there simply are not as many firms carrying it. Careful reading of advertising will show you who sells what. There are some commercial and industrial suppliers who carry the units at prices only slightly higher than mail order surplus dealers. Surplus prices are all about the same, but since slight discrepancies do exist, shop around before you send anyone your money.

There are quite a few regulators available that have not been mentioned here. Don't be afraid to try some of them; however, remember to make sure that you are buying a fixed voltage, three-terminal voltage regulator integrated circuit. Some chips, such as the 723, are not three-terminal devices, and they require a number of external components to adjust the output voltage and set the current limiting point. They have to be used with a much larger number of external components, in circuits whose design is much more involved than this. The giveaway should be that no specific output voltage is mentioned. If the device is advertised simply as a monolithic voltage regulator, you can be pretty sure that it is a multiterminal unit. Some ads do not give a verbal description of the device, but just its number and price. You can cross-reference to other advertisements to figure out what the product is.

A series of regulators that you might be interested in, especially if you want to build a ±15V op amp supply, is the µA7900 series. They are negative voltage complements to the µA7800s. Availability is still somewhat limited, but hopefully it will improve shortly. Both series are produced in a wide range of output voltages, including a couple that have not been mentioned here. Again, check through the ads in the back of magazines.

Make sure that, for any device you buy, you also obtain a set of the manufacturer's specification sheets and also, if possible, application notes. Originally the LM309 was produced by National Semiconductor and the µA7800 series by Fairchild, but quite a number of other manufacturers have since second-sourced them. So don't be too upset if you receive from a single distributor spec sheets and regulators with different manufacturers' names on them. There is enough industry-wide compatibility so that you need not worry about differences between manufacturers. (Actually, a more likely case would be to find no manufacturer's imprint at all on one or the other.) Hopefully, a little bit of reading through those data sheets and application notes will answer most of your questions.

There are external circuits, available from suppliers' and manufacturers' application notes, that can be used to make these regulators even more versatile. A typical circuit in this category is one enabling the output voltage to be set at different points, or even to be continuously adjustable.

Three-terminal regulators come into their own for the simple supplies. Usually, by the time you hang on all of the necessary parts to one of these three-terminal units to achieve a specialized application, you find that you would have been better off by using a multiterminal device from the start. The 723 mentioned before is only one of a large number of these chips available.

SIMPLE 5V SUPPLY

As shown in Fig. 2-31, the supply consists of a simple bridge rectifier, fairly heavy filtering with two 2000 µF elec-

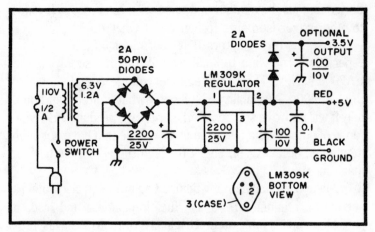

Fig. 2-31. Schematic of simple 5V supply with 3.5V option.

trolytics, and regulation with an LM309K integrated circuit regulator. As shown in the photos, the LM309K is mounted on an aluminum heat sink right on the case (the IC case is grounded so no mica insulators are needed). With such a heat sink the circuit will provide up to 1A at 5V. With the IC mounted just on the case, but without a separate heat sink with fins, the maximum current output will be somewhat less, but probably still above 0.5A.

The secret of the circuit, of course, is the regulator IC. Not only does it provide excellent regulation and practically eliminate any ripple on the output, but it is also short-circuit proof; you can short the output of the supply and no harm will be done. It also shuts itself off in case its temperature gets too high; in other words, no damage will be done if you skimp on the heat sink. You just get less output current before it shuts off.

The 5V output is perfect for 7400-series and other TTL ICs, as well as most DTL and ECL circuits. Since RTL ICs need only about 3.6V, you can add an optional output for these ICs by just adding two diodes and one more electrolytic.

To reduce power supply feedthrough of logic signals from one circuit to another, place the 100 μF and 0.01 μF capacitors as close as possible to the two output terminals, and use short leads. In addition, don't forget that TTL ICs gener-

162

ate very short spikes on their power supply lines; therefore, you will need additional bypassing inside your circuitry itself. Every four or five ICs should have their own 0.1 or 0.01 μF disk capacitor connected directly from the +5V line to the nearest ground, using the shortest possible leads. These additional capacitors should be as close to the ICs being filtered as possible. In really severe cases, you may have to connect these capacitors right at each IC.

Chances are good that you won't need a bigger supply than this; it has been used to power a frequency counter with over 30 TTL ICs which only took slightly more than 0.5A. Not many projects will ever need more than 1A. But if you do need a bigger supply, don't try paralleling the LM309K regulators to increase their current capacity, since they will not share the current equally. The best bet would be to make several separate supplies, which share only a common transformer, rectifier, and brute-force filter (at least 4000 μF per ampere). Each output would then be separately regulated, and would drive a separate part of the circuit.

PRECISION 10.000V DC REFERENCE

This circuit is a simple, easy to build DC voltage reference standard, whose temperature coefficient can be tailored to individual requirements. It can be used by itself or as an individual circuit element.

Design Considerations

The circuit described (Fig. 2-32) uses an operational amplifier in a noninverting circuit utilizing a single positive supply. The output of the reference standard supplies the zener reference diode current, providing a stable supply voltage for the zener and decoupling from the positive supply voltage. The positive supply voltage should be regulated by either a three-terminal regulator or zener diode and should be from +5V to +18V.

Temperature stability is expressed in percent per degrees Celsius or ppm (parts per million) per degrees Celsius (0.0001%/°C is equivalent to 1 ppm/°C). A 1 ppm/°C change referred to an output of 10.000 volts is 10 μV/°C.

Zener reference diode CR1 should not be confused with the more common zener regulator diode. The zener reference diode is intended for use in applications where it is important to maintain stable DC voltages under severe combinations of temperature, shock, and vibration. The temperature stability of the zener reference diode is due in part to the combination use of reverse-biased and forward-biased silicon PN junctions, taking advantage of their opposing temperature coefficient characteristics. Application notes and design data sheets are available from the larger suppliers of zener reference diodes and are invaluable as reference material.

The 1N821 through 1N829 family of zener reference diodes is used for CR1. The diodes in this family exhibit temperature coefficients from 0.01%/°C (1N821) to 0.0005%/C (1N829) at a nominal zener current of 7.5 mA. The nominal zener voltage is 6.2V ±5%. R1, which sets the nominal zener current (7.5 mA in this case), should be a 100 ppm/°C metal film resistor. Other families of zener reference diodes can be used, such as the 1N4565 through 1N4584, with a corresponding change in nominal voltage and current.

As a general rule, the gain resistors used in the reference standard should have temperature coefficients similar to the zener reference diode selected. For example, if the 1N821 is selected, gain resistors R2 and R3 should have temperature coefficients of 0.01%/°C (100 ppm/°C).

The overall temperature coefficient of the reference standard is dependent upon those of the reference zener CR1, gain resistors R2 and R3, and to some extent, zener current resistor R1 (providing it has a 100 ppm/°C temperature coefficient) and A1.

The gain required of the circuit is dependent upon the zener voltage. The nominal voltage of the 1N821 is 6.2V, necessitating a gain of 1.613 for an output of +10.000V.

The total resistance of R2 and R3 should be 5K to 10K limiting the current through the gain resistors to from 1 mA to 2 mA.

If R1 is a 100 ppm/°C type as suggested, here is an easy way to estimate the overall temperature coefficient of the

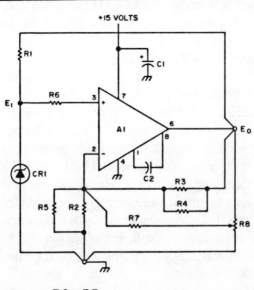

$$Gain = \frac{R2 + R3}{R2}$$

$$E_0 = E_1 \frac{R2+R3}{R2}$$

A1 — LM301A Operational Amplifier

C1 — 1 mF electrolytic capacitor, 25 W V dc

C2 — 30 pF ceramic disc capacitor

CR1 — Reference zener diode, 1N821 family

R1 — 100 ppm/°C metal film resistor (Select value for the nominal current of CR1.)

R2,R3 — Gain resistors (Select for desired tempco.)

R4,R5 — Gain trim resistors

R6 — ½ Watt resistor (Value should be the parallel equivalent of R2 and R3.)

R7* — Optional +10.000 volt trim resistor for use with R8 (Value can be 100k to 1 megohm, 5-10 ppm/°C.)

R8* — 10k 89PR cermet pot or equivalent optional +10.000 adjust.

*with a 100k resistor at R7, the output can be adjusted about 10% if the sum of R2 and R3 is 10k. 1 megohm will provide about 1%.

Fig. 2-32. Schematic of precision 10.000V supply with parts list.

165

completed reference standard: (1) write down the temperature coefficients of CR1, R2, and R3 in ppm per degrees Celsius, (2) square each one that you have written down, (3) add the squares of R2 and R3 together, dividing the answer by seven, (4) add the square of CR1 and the answer of step 3, and (5) find the square root of the answer in step 4. Step 5 will be the approximate overall temperature coefficient of the reference standard in ppm per degrees Celsius. This will give you a good idea if the components you have selected fit your temperature coefficient requirements. If CR1, R2, and R3 have the same temperature coefficient just multiply *one* of them by 1.1 to detemine the overall temperature coefficient of the reference standard. The reason for the division by seven in step 3 is that the temperature coefficient of R2 and R3 is reduced a factor of approximately 2.7 (depending on the gain) by the low gain of the amplifier.

The output of the reference standard can be trimmed to +10.000V by paralleling a resistor across R2 or R3 as required. Temperature coefficient of gain trim resistors R4 and R5 is critical and should not alter the tempco of the gain resistors. For example, if R2 was a 6.04K 25 ppm/°C resistor and a 62K ½W carbon resistor was paralleled across it having a 1000 ppm/°C temperature coefficient (typical of carbon resistors), the result would be the same as using a 5.5K 100 ppm/°C resistor. The effective temperature coefficient of the trim resistor in this case would be a tenth its real value. A 620K ½W carbon, in the same example, would have an effective temperature coefficient of a one-hundredth or 10 ppm/°C, and a 6.2M would have an effective temperature coefficient of a one-thousandth of or 1 ppm/°C. Some discretion must be used when selecting the value and temperature coefficient of the trim resistors.

Practical Design

The next step is to build a working reference standard. The following components were selected for a reference standard having a calculated temperature coefficient of 22 ppm/°C and a measured temperature coefficient of 17 ppm/°C after assembly. An LM301A operational amplifier was selected for A1. The LM301A will supply approximately 20 mA and has a typical temperature coefficient of 1 ppm/°C. A 1N825 reference zener was selected for CR1 which has a 20 ppm/°C temperature coefficient and a nominal voltage of 6.2V at 7.5 mA. An RN55 100 ppm/°C 511Ω metal film resistor was selected for R1. The actual reference zener voltage at 7.5 mA measured 6.286V, necessitating a gain of 1.591 for +10.000V output. For a gain of 1.591, an RN55E 25 ppm/°C 6.04K metal film resistor was selected for R2 and an RN55E 25 ppm/°C 3.57K metal film selected for R3. A ½W 2.2K 5% carbon resistor was selected for R6. After assembly, R3 had to be trimmed for an output of +10.000V.

The most difficult task after assembly will be finding access to a 4 or 5 digit digital voltmeter so that the reference standard output can be trimmed to +10.000V. No special equipment is required up to the trimming operation. The accuracy of the reference standard will be as good as the equipment used to trim it.

Conclusion

It has been demonstrated that a precision reference standard can be constructed by utilizing available reference zener diodes and suitable gain resistors. This circuit is primarily a tool to enable some of those interested in acquiring a stable DC source to design their own. Different output voltages can be obtained by either changing the gain of the amplifier or adding a voltage divider across the output of the reference standard (current cannot be supplied in this mode). Layout is not critical, cost is minimal, and performance can be determined to some extent before assembly.

SIMPLE FOUR-VOLTAGE BENCH SUPPLY

This power supply provides the choice of four popular output voltages, three current overload limits, and an overload indicator, yet requires very few components. This principle of current overload protection can be easily adapted to existing power supplies. The size of heat sink required for a

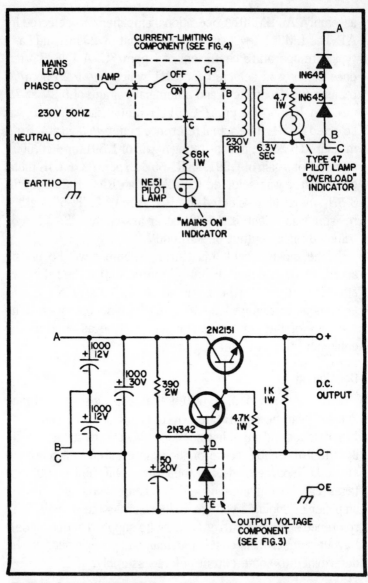

Fig. 2-33. Schematic of simple four-voltage bench supply. The components in the dashed squares are simplifed examples for the purposes of explanation. A 230V supply is shown but the system can be easily adapted for 110V as described in the text. The overload indicator turns off when excessive currents are drawn.

series regulator transistor can be reduced if this overload principle is used. Specifications for this supply are:

168

Output voltage .... 3, 6, 9, or 12V (switched)
Output current .... 200 mA maximum
Internal impedance .... Less than 5Ω, preferably about 1Ω.
Ripple .... Less than 10 mV P-P
Overload Protection .... To reduce the output voltage to a
low level in the event of exces-
sive load currents or short-
circuits on the output terminals.

The circuit diagram is shown in Fig. 2-33 and the DC output characteristics shown in Fig. 2-34. Note that the input supply voltage is 230V at 50 Hz. This is the domestic supply used in New Zealand and has been retained in this circuit in order to show the overload protection arrangements. Some component values will have to be changed when 110V primary transformers are used. As access to 110V 60 Hz systems are

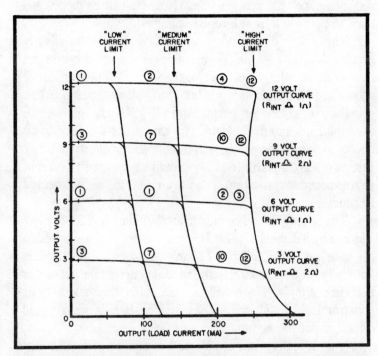

Fig. 2-34. Output characteristics of the simple four-voltage bench supply. Note that the four voltages—3V, 6V, 9V, and 12V— and three current limits—50 to 100 mA, 130 mA, and 240 mA—are available. The maximum current limit may be increased (see text). The figures in circles represent peak-to-peak ripple voltages (in millivolts) at the load points shown measured with an oscilloscope. Output impedance is about 2Ω at worst on the flat part of the curves.

just about impossible in ZL-land, details will be given later on the methods used to select component values for other systems of power distribution.

The Circuit

The regulator circuit is basically a transistor series regulator with a zener diode as voltage reference. Two transistors are used in a compound circuit as cascaded emitter followers to ensure a low value of output impedance and to prevent any wide load variations from causing excessive changes to the zener diode operating conditions. A zener diode reference voltage is selected by means of switches to provide the output voltage required. The switching used is shown in Fig. 2-35. Two toggle or slide switches select the required zener reference diodes. Note that three diodes are used to provide four reference voltages.

The zener diode reference voltage has to be slightly greater than the resulting power supply output to allow for the base-emitter voltages of the two transistors. The transistors shown are NPN silicon types but many other types could be substituted, silicon or germanium. If PNP transistors are used, all diodes and all electrolytic capacitors will have to be reversed to provide voltages of the opposite polarity. If other transistors are substituted they should have as high a current gain as possible consistent with the current they are expected to handle.

The supply should be built in an aluminum box 5 inches by 4 inches by 3 inches. The 2N2151 power transistor is bolted to the case but insulated from it by mica washers. A heat sink is not really required owing to the modest current output of this particular supply. More will be said about this later. Other constructional details will not be given for there is nothing at all critical about the layout.

The 390Ω resistor is selected to ensure that the zener diodes do not dissipate more than their 400 mW under any conditions. In fact they are considerably underrun. Other zener types could be substituted to provide other output voltages if required, but if this is done then some adjustment of

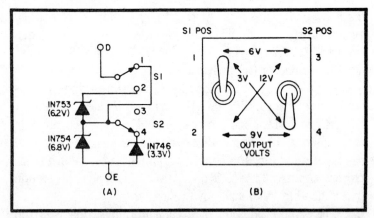

Fig. 2-35. Current-limiting switch arrangement. Two SPDT switches provide off and choice of three output current limits by selecting one of two capacitors or both in series. The neon bulb operates whenever the supply is on; it doesn't matter which current limit range is used.

the 390Ω resistor may be necessary. The zeners can be individually selected if enough are available, to provide accurate output voltages.

The voltage doubler is conventional and again a wide range of diodes could be used in place of the ones shown. A pilot light using a type 47 lamp is connected to the secondary of the transformer. A 4.7 Ω resistor is used to protect the lamp from the rise in secondary voltage in a no-load state but this will be discussed later.

Overload Protection

The regulator system is capable of regulating the output against wide changes of input voltage from the main supply, and it is this characteristic that is used in this unit to give overload protection. From the secondary of the transformer to the output, the regulator is of quite common design. Such a regulator will normally regulate up to the rated current limit of the series regulator transistor and other components, but will have little or no immunity against overload. Excessive load currents will generally cause overheating in the regulator with ultimate failure of one or several components.

Many different techniques have been devised to protect this type of supply from overload. The most elaborate methods use some form of trigger circuit to electronically switch off the supply after a certain load current has been reached. This entails a reset procedure to restore the output after the cause of the overload has been removed.

Another type of overload system ensures that the output voltage is abruptly reduced to a low and safe level after a certain load current has been reached. This type of protection is usually self-resetting in that once the cause of the overload has been removed, the supply will immediately return to its normal output voltage. It is one form of this type of protection that is used in this small power supply.

The simple addition of a capacitor in series with the transformer primary provides a convenient current overload protection with some other advantages. This capacitor is shown as C_p in Fig. 2-36. The effect is to cause the vertical voltage drop as shown on the characteristics in Fig. 2-34. It may seem at first to be a highly unlikely result so an outline of the action will soon be given. The high-current overload does double back as shown on Fig. 2-34.

The switching arrangements for the current overload are shown in Fig. 2-35. Two SPDT toggle or slide switches (similar to the zener-switching ones) provide a mains-off facility and the choice of three overload current limits. Two capacitors are used, and these are connected individually and in series to provide a range of three capacitance values.

The Indicators

The neon pilot lamp is used to indicate that the supply is on. In one set of switch positions the neon is fed with 230V through the 0.5 μF capacitor. This capacitor has a reactance of 6400Ω at 50 Hz so has negligible effect when in series with the neon current-limiting resistance. This form of connection is necessary to ensure that the neon glows when the switches are in all three current-limiting positions, and yet is extinguished when the switches are in the mains-off position.

It may seem strange that two pilot lights are necessary. The neon shows when the power is on while the other lamp shows when an overload has occurred. Type 47 lamp extin-

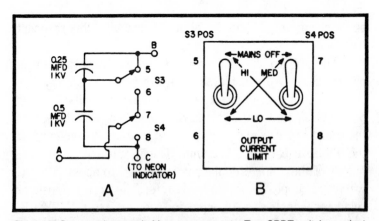

Fig. 2-36. Output voltage switching arrangements. Two SPDT switches select the required zener diode. Switch positions in portion A are also shown in portion B, which illustrates the panel labeling. Note that for 3V as shown the 3V zener is in parallel with a 6V zener, which of course will not be operating.

guishes when an overload has been applied. While both lamps are lighted the supply is delivering its rated voltage. If the type 47 goes out then an overload has occurred. This arrangement works irrespective of the voltage output setting or the current overload range that is being used. With no voltmeters or ammeters used in the output (they were not considered necessary for the application in which this supply is used) this twin-lamp system has proved to be quite effective.

Overload Circuit Operation

The action of the series capacitor in the primary can be explained when it is realized that the input impedance of the primary of the transformer (when the DC output is on open-circuit) can be regarded in its simplest form as an inductance in series with a resistance. The inductance value is made up of transformer leakage reactance and some primary inductance, while the resistance is made up of resistances representing the transformer losses and the load presented by the regulator components. The primary circuit is thus a series-tuned circuit, consisting of C_p, and the transformer inductance and resistance. This tuned circuit must not be at resonance at the mains frequency or excessive primary current would flow. The resonant frequency is made higher than the mains frequency, and a change of C_p has the effect of changing this resonant frequency.

The primary current (with no load on the output terminals) is increased considerably with the inclusion of C_p. The current flowing is not in phase with the supply voltage so the power dissipated in the primary is small. The voltage across the transformer primary is increased to about 300 volts when the 0.5 μF capacitor is in circuit. This in turn means that the output of the voltage doubler is increased to about 20V DC when the supply is in a no-load state. The series transistor regulator accepts this 20V input and reduces it to 12V, 9V, 6V, or 3V at the DC output terminals, even in a no-load state.

When a load is connected to the regulator DC output terminals, it has the effect of changing both the inductance and resistance values seen at the transformer primary. The resistance value increases as the load increases, while the inductance value decreases. The resonant frequency of the primary tuned circuit, therefore, rises and the impedance presented to the mains input increases, causing a decrease in primary current. The Q of this series tuned circuit thus falls. The output of the voltage doubler and the 20V DC level at the input to the regulator drops.

At some value of load current, this DC voltage will have dropped to a point where the regulator can no longer operate satisfactorily, and the output DC voltage now takes a downward plunge. The exact mechanism causing this steep fall is rather complex for the current and voltage waveforms at the transformer primary become distorted even though core saturation has not been reached.

At the overload point at the top of the downward plunge of the voltage curve, the type 47 pilot lamp goes out, showing that the transformer secondary voltage has dropped to a low level. With an overload current at the DC output terminals, the voltage at the output of the voltage doubler has dropped to about the same voltage as that at the DC output terminals, and with a short circuit, it drops to zero.

The overall result is most effective and especially is this so when components economy is considered.

The voltage rating of the capacitors must be high because under no-load conditions the voltage across the capacitor rises to about twice the mains voltage. The voltage rating should be 1 kV for 230V systems and 500 V for 110V systems to be sure that the component will not fail.

Determining Component Values

This current-limiting technique has been applied to another power supply and it was found that a 4 μF series capacitor (1 kV, oil-filled) produced as steep a falloff as the 0.5 μF on the supply described. The value of capacitor to use for a particular value of current limiting is just about impossible to calculate because little is usually known about the transformer input impedance for the resulting DC load. The capacitor value is best determined experimentally by substituting value after value (starting off with a 0.1 μF for 230V systems and a 0.25 μF for 110V systems) and plotting the output characteristics using an ammeter voltmeter, and adjustable load resistor of suitable value. Increase the capacitor values in, say, 0.25 μF steps until the desired cutoff current is reached.

One point must not be overlooked. If you try this with an existing supply make sure that the voltage ratings of any filter capacitors are not exceeded. Remember that the DC voltage output of the rectifier is increased by approximately 50% when the supply is in a no-load condition. The primary voltage is increased by a like amount, but this is not as serious as it appears to be for the power factor is far from unity and little increase in transformer operating temperature (if any) will result. An AC voltmeter across the primary can be used to keep this under observation while substituting capacitor values. Remember that excessive primary currents can be drawn if the primary approaches resonance, so start off with the values of capacity quoted and increase it in small increments until the desired limit is reached, keeping an eye on filter capacitor voltage ratings. Remember too that you are playing with mains power so play it safe.

Other Advantages

Two other unexpected benefits also accrue from this series capacitor current-limiting arrangement. When first switching on, all the filter capacitors are discharged and the diodes would normally pass a very high current during charging. With this series capacitor in circuit, this charging current appears as an overload current, and the initial charging current is limited. A resistor in series with the transformer secondary to limit the switch-on charging current as often used is now not necessary.

The second added advantage concerns the size of the power transistor heat sink (if used). Because the DC input voltage to the regulator falls as the load current increases, the power dissipated in the series regulator is not as high as would normally be expected, and large DC output currents can be handled before a heat sink becomes necessary. This alone is an economy in space.

A change in mains frequency will have the effect of changing the current-limit level but this is unlikely to be bothersome if you draw your power from a large distribution system. A change of input mains voltage has a very small effect on the limit level and can be overlooked.

Ripple voltages, measured with a Tektronix 545B oscilloscope, are shown in circles in Fig. 3-24 (millivolt peak-to-peak values) at various load levels. The lower output impedance and inferior ripple of the 3V and 9V curves can be attributed to the use of the 3V zener with its high dynamic impedance.

HEFTY 12V SUPPLY

If you've ever needed a good, regulated 12V power supply to sit on your test bench, this is the one. It provides an adjustable output over an 11V to 14V range, keeping regulation to better than 100 mV. The output is set to current limit at 6.5 A, although you can change that to 13A. Additionally, the power supply is protected against transients and reverse voltages at the output terminals.

Figure 2-37 shows a conventional regulated high current power supply. Using a 723 voltage regulator gives the necessary precision and regulation; a hefty external pass transistor, complementing the 723's internal pass transistor, provides the power. Although this type of circuit works reasonably

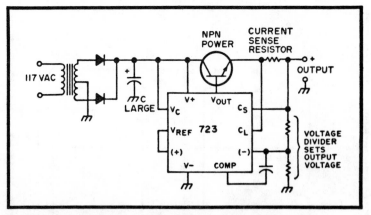

Fig. 2-37. Schematic of conventional regulated high current supply.

well, making a few circuit changes can achieve far better performance.

Modification number one is to use a Darlington power device as the pass element. The MJ3000 specified has a gain of 1000 at 5A. This high gain requires less base current, keeping a light load on the regulator. The actual amount required to drive the MJ3000 is about 1.5 mA per ampere of output current. Taking the 723 quiescent current into account, the 723 is left with a worst-case current load of under 12 mA, promoting a stable and cool-running regulator section.

To understand modification number two it's necessary to look at the schematic of Fig. 2-38 and fool around with some numbers.

With a conventional capacitive input filter (which this is) ripple voltage increases linearly with output current. Despite the large amount of capacitance (five paralleled 4000 μF capacitor for a total of 20,000 μF), in a situation where you're drawing 6.5A there is a ripple voltage of about 3V peak to peak across the filter bank. Like all IC regulators, the 723 must have a somewhat higher voltage at its input than at its output in order to maintain regulation (typically 5V). If we were to take the input to the 723 from the filter capacitor output, as in the typical configuration of Fig. 2-37 there may not be enough drive to maintain regulation. Hence, in Fig. 2-38, a couple of diodes and another filter capacitor derive the input voltage

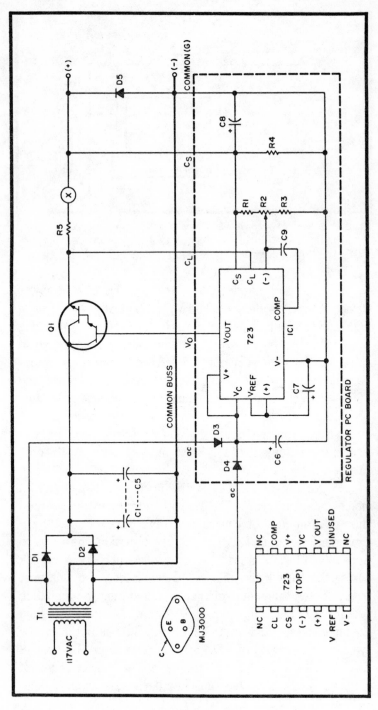

Fig. 2-38. Modified high current supply using the MJ3000 Darlington.

178

directly from the transformer secondary. Due to the light load current on the 723 discussed earlier, we can get away with a 250 μF capacitor; at 6.5A of output current, this means a ripple of 300 mV peak to peak at the regulator input, which is certainly enough to keep the 723 happy and regulating.

A few other points are worth mentioning. Bypassing V_{ref} to ground with a small tantalum capacitor substantially reduces any noise and ripple appearing at its output. Another point to consider is temperature stability (i.e., constant output voltage regardless of temperature). Resistors R1, R2, and R3 set the output voltage. Since we're using carbon composition resistors, you might think the stability would not be too good; however, what matters in this case is the resistance ratio rather than absolute values. Since for 12V operation the trimmer control is in its approximate midrange position, any temp-

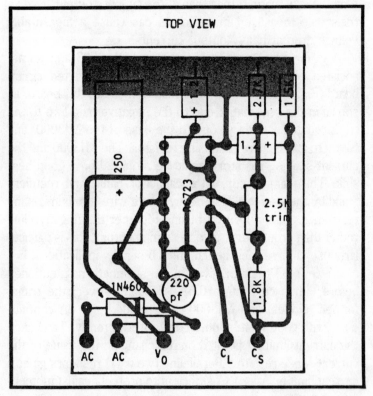

Fig. 2-39. PC board and parts layout for the 12V hefty supply.

erature variations tend to change all resistances equally, pre serving the proper ratio and giving a constant output voltage. One extra point to consider is by tagging a resistor on the output to give a small minimum load, the regulator is always in its active state, maintaining better stability and accuracy. The tantalum capacitor connected across the output improves transient response.

Finally, at the extreme output of the power supply is a diode. Note that it's connected up to shunt any negative spikes or transients to ground before they can get into the power supply to cause trouble.

One protection device that is not necessary but highly desirable is a fuse; in case anything should go wrong with the supply, it protects the device being powered. Because a fuse has a small amount of resistance, it should be inserted at point X on the schematic rather than at the supply's output. Even a resistance measured in milliohms can cause a measurable voltage drop at heavy output currents.

As far as assembly is concerned, the 723 and its associated components mount on a separate printed circuit board (Fig. 2-39). Six connections are made to this board, as shown in the schematic: one to the negative bus, two to the transformer secondary, one to the base of the MJ3000, and two across the current sensing resistor. The MJ3000 and the current sensing resistor should be mounted on a good heat sink. The transformer, filter capacitor bank, and rectifiers stand by themselves. Use heavy gauge wire to minimize any resistance. Also note that busing together the negative line rather than grounding it gives a floating output; this is particularly useful in service bench and lab supply applications.

The 0.1Ω resistor given in the schematic as mentioned before, limits current to 6.5A, ±5%; however, the transformer diodes, and MJ3000 can provide up to 8A continuously with good ventilation, or 10A intermittently. To change the current limiting point requires changing the value of the current sense resistor. Paralleling two 0.1Ω resistors gives a current limit point of 13A sufficient to protect against momentary shorts.

Using these same principles can give power supplies capable of delivering up to 100A of regulated current. All that is required is making a more powerful transformer-diode-heat sink combination, increasing the filter capacitance, and paralleling an appropriate number of Darlington devices, adding a small value resistor on the emitter of each Darlington.

Parts List

R1 ....1.8K ¼W 10%
R2...2.5K pot
R3 ....2.7K ¼W 10%
R4 ....1.5K ¼W 10%
R5 ....0.1Ω 5W or greater (see text)
C1-C5 ....4000 μF at 20V
C6 ....250 μF at 25V
C7, C8 ....1.2 μF at 35V
C9...220 pF at 25V
D1, D2......................Rectifier diode at 6A (Motorola MR1120)
D3, D4...1N4607
D5...1N4002
Q1... ....MJ3000 (Motorola) Darlington power transistor
IC1...LM723 integrated circuit
T1.........................24 − 28V center tapped transformer at 4A
Misc ....Line cord, chassis, mica insulators plus
                          silicone grease, heavy gauge wire, fuse
                          and fuse post, binding posts, etc.

78MG & 79MG VOLTAGE REGULATORS

Several manufacturers have offered three-terminal fixed voltage IC regulators for some time. These ICs just about took all of the work out of power supply design, as long as you only wanted one fixed power supply voltage; however, making up a variable supply was a little messy.

Well, designers at Fairchild Semiconductor realized many IC users require voltages other than the ones that they had standardized for their series of fixed voltage regulators. So they introduced two ICs that fill in the gaps in their series of fixed voltage regulators. These two circuits are the 78MG positive regulator and the 79MG negative regulator.

Here basic circuits are discussed for using these two ICs at fixed voltage levels. As with most ICs it doesn't do to simplify, so we will also show some of the pitfalls to avoid.

Features

The 78MG and 79MG variable output voltage regulators offer unique features which make the circuits very simple to use, while providing protective features which are hard to obtain in a discrete design.

The most outstanding characteristics of these IC regulators are protection against output short circuits and protection against excessive power dissipation. These features make the regulators difficult if not impossible to destroy through misuse. A third related characteristic limits the ICs' output transistors to safe area operation.

The regulators are designed to supply up to 500 mA of current, with input voltages in the range of +5V to +35V for the 78MG, and −2.2 to −30V for the 79MG. The maximum input voltage for either is 40V.

The devices are packaged in what is called a mini-botwing four-terminal package. The wings on the package allow the device to be secured to a heat sink to increase power dissipation capabilities. Both devices feature line regulation of 1% when the input voltage is varied from 7V to 35V and load regulation of 2% when the load current is varied from 2 mA to 500 mA. The quiescent current is 2.5 mA, and the regulators require a minimum of 2V margin between input and output to stay in regulation. The power dissipation is internally limited to 7.5W.

The older three-terminal 78 series fixed voltage regulators have an internal voltage reference which is sensed through a resistive voltage divider network. The ratio of resistance in the voltage divider network is preset internally when the chip is manufactured. The 78MG and the 79MG regulators use the same concept, except that there is no internal voltage divider network. Instead, this point is brought outside the package by means of a fourth pin. Thus, all you have to do is to set the regulator for any voltage within its range. This is all that is necessary to supply two resistors of the proper value to supply the feedback voltage to this fourth terminal.

The package is basically a stretched out four-pin plastic

dip package with two wings protruding from each side for a heat sink.

The Insides

Figure 2-40 shows a block diagram of the 78MG and 79MG regulators. The internal structure, with a few minor refinements, is identical to that of the three-terminal regulators. The major exception is that one input to the error amplifier is brought outside the package so that you can supply any reference voltage you wish.

Let's get a brief idea of internal circuit operation before getting caught up in applications. The startup circuit contains a zener diode and two transistors, which have the purpose of bringing the circuit into initial regulation. After it is in regulation, the startup circuit is biased off. The error amplifier compares the voltage at the control input to the internal voltage reference and generates an error signal of the proper polarity if the two voltages are not the same. This error signal either increases or decreases the bias current into the series pass transistor, which in turn increases or decreases the regulated output.

The safe area limit and thermal shut-down circuits provide protection against normal operating overloads. The short-circuit protection reduces the current available to the series pass transistor in the event of an output short circuit.

The regulators also contain a 30 pF MOS capacitor to increase stability and lessen the possibility of oscillation. The regulators achieve thermal stability through careful balancing of positive and negative temperature-coefficient components.

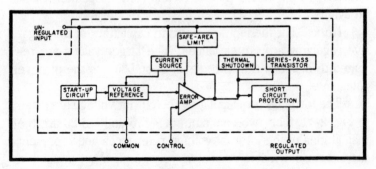

Fig. 2-40. Block diagram of the 78MG and 79MG regulators.

Basic Regulator

Figure 2-41 shows the simplest configuration in which the positive and negative regulator can be used. This is really all you need to place in front of an unregulated power source in order to produce a stable well-regulated voltage source.

Input capacitor C1 is necessary to stabilize the regulator under all possible conditions. Although the manufacturer lists this input capacitor as an optional item, needed only if the regulator is located more than a few inches from the unregulated supply filter capacitor, it is strongly suggested to always include it. For the few pennies of extra cost, you may save yourself from destroying a regulator chip. It also allows you a lot more freedom in locating the unregulated source, and in choosing the unregulated supply filter capacitor.

Conversely, C2, the output capacitor, is really not necessary for regulator operation, although it does tend to improve transient response. If you intend to use the regulator to supply power to switching loads, such as TTL circuits, you would be well advised to include an output capacitor of approximately the value shown. One further comment about output capacitance: if there is any chance that you might drive a very large capacitive load (e.g., as large as the unregulated supply filter capacitor), you must provide extra protection for the regulator to prevent possible destruction of the chip. A diode in series with the output, large enough to handle all the output current, will provide quite adequate protection from reverse voltage. In all cases, the application of a reverse voltage between the output and input pins of the 78MG and 79MG regulators must be avoided.

Notice that the input and output capacitors of the negative regulator are larger values. The larger values are required to maintain the same level of stability as the positive regulator.

The formulas for calculating the output voltage are given in Fig. 2-41. The two constants of $+5.0$ and -2.23 are the magnitudes of the internal voltage references and represent the lowest obtainable regulated voltage. If a pot is used in place of R1 and R2, an additional fixed resistor should be

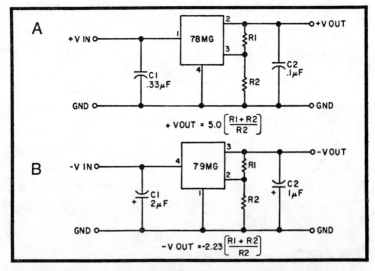

Fig. 2-41. Basic regulator circuits. (A) Positive regulator. (B) Negative regulator.

included in the ground leg of the pot. This linearizes the relationship between the rotor position of the pot and the output voltage. In fact, if you use a fairly accurate linear taper pot, there is no need for an output voltmeter; just mark off the pot rotation in equal increments, and you have a nice permanently calibrated variable voltage source.

Every regulator needs an unregulated supply. Figure 2-42 shows a supply suitable for most general purpose work Values are shown in Fig. 2-43 for two common voltage ranges: As a rule of thumb, be generous with voltage ratings on large filter capacitors. Not only are the large-valued capacitors more susceptible to failure near their rated voltages, but many inexpensive, low voltage transformers greatly exceed their nominal output voltage at low current levels.

Another important item to keep in mind is that the safe area limiting circuit of the voltage regulator IC begins to operate when the voltage across the series pass transistor exceeds 8V. This circuit interacts with the short-circuit protection transistor to limit output current. Thus you may not be able to take advantage of all the available output current rating of the regulator if the differential between unregulated input and regulated output voltages is too great.

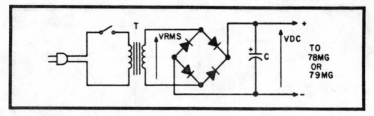

Fig. 2-42. Unregulated power supply.

Another important point to keep in mind is that, for proper regulation, there must be a minimum of 2V differential across the IC regulator. This, of course, means that your unregulated supply must provide at least 2V more than your desired output voltage.

Other Circuits

A 500 mA regulator is fine for a small supply for most projects; however, it's not nearly enough for a main lab power supply. What's needed is more like 1A to 2A of output current.

Increasing the current output of the 78MG and the 79MG regulators can be done simply by adding an external series pass transistor and a biasing resistor. This hookup is shown in Fig. 2-44 for currents of about 1.5A. This circuit, however, provides no short-circuit protection or safe area limiting for

Vrms NOMINAL TRANSFORMER SECONDARY VOLTAGE (Vrms)	Vdc AVERAGE DC OUTPUT VOLTAGE (Vdc)	Vout DESIRED REGULATED OUTPUT VOLTAGE (Vdc)	I1 MAX. LOAD CURRENT (Amps)	C1 MIN. CAPACITOR SIZE (Microfarads)
6.3	8.2	2.23* to 5.3	0.5	1200 @ 20 V
		2.23* to 5.3	1.5	3600 @ 20 V
		2.23 to 4.5*	0.5	600 @ 20 V
		2.23 to 4.5*	1.5	1800 @ 20 V
12.6	16.3	2.23* to 12.5	0.5**	600 @ 35 V
		2.23* to 12.5	1.5	2000 @ 35 V
		2.23* to 11	0.5**	300 @ 35 V
		2.23* to 11	1.5	1000 @ 35 V
25.2	32.8	5 to 29.5	0.5**	300 @ 75 V
		5 to 29.5	1.5	1000 @ 75 V
		5 to 26	0.5**	150 @ 75 V
		5 to 26	1.5	500 @ 75 V

*This voltage is obtained only with 79MG.
**If the 78MG/79MG is used without an external pass transistor, full current may not be available for lower output voltages, since the IC will limit power dissipation to 7.5 Watts.

Fig. 2-43. Component values for the unregulated power supply shown in Fig. 2-42.

186

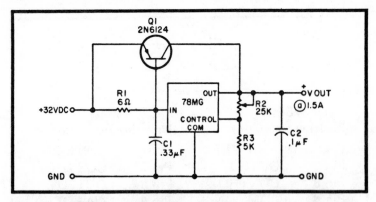

Fig. 2-44. Adjustable voltage regulator with an output of 5V to 30V at 1.5A. Transistor Q1 should have a large heat sink, preferably in common with the 78MG regulator; however, be sure to electrically isolate the 78MG and Q1.

the external pass transistor. Thermal overload protection of a somewhat lesser degree can be provided by mounting the pass transistor and the 78MG in close proximity on the same heat sink.

To regain short-circuit protection for the external transistor, an additional small signal transistor and current-sensing resistor are required. Use the circuit shown in Fig. 2-45. This connection uses an NPN transistor for the external series pass. The collector of Q1 can be tied either to the unregulated input to the 78MG, or to some other positive voltage. This means you can use the 78MG to control a much larger voltage than it could ordinarily handle. The limiting factor then becomes the breakdown voltage of Q1.

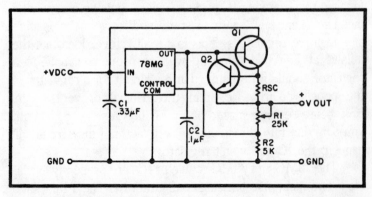

Fig. 2-45. High current regulator has short-circuit protection set by the value of Rsc.

187

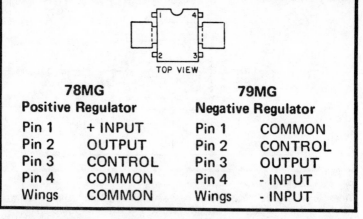

TOP VIEW

78MG Positive Regulator		79MG Negative Regulator	
Pin 1	+ INPUT	Pin 1	COMMON
Pin 2	OUTPUT	Pin 2	CONTROL
Pin 3	CONTROL	Pin 3	OUTPUT
Pin 4	COMMON	Pin 4	- INPUT
Wings	COMMON	Wings	- INPUT

Fig. 2-46. Regulator pin designations. It is a good idea to electrically isolate the bat wings from the heat sink. The wings are internally connected to pin 4 of the package, which means that they are at the same potential as ground in the 78MG, but at the input potential in the 79MG.

Since the output circuit is proportional to the voltage across Rsc, that voltage may be used to indicate output current. A simple metering circuit can provide a measurement of the load current. Of course, current will be drawn by R1 and R2 even when there is no load current being drawn; however, this current should be negligible in comparison to the normal supply current used.

Pin Connections

Figure 2-46 shows the terminal connections for the four-pin mini-batwing IC package. Note that the functions of pins 1 and 4 and pins 2 and 3 are reversed between the positive and negative regulators. This is a really clever idea that the fellows at Fairchild came up with to prevent damage to a regulator should it be plugged into a circuit for a regulator of the opposite polarity. Reversing the terminals prevents the IC from being reverse biased and probably destroyed. Thus if you plug one into the wrong socket, nothing disastrous will happen; the IC just won't regulate.

The 78MG and 79MG regulator ICs are presently available in two main mini-batwing configurations. The T1 package has the batwings bent downward so that they can be inserted into a PC board slot and soldered down. The T2 package has the leads bent slightly downward and then out horizontally, so the wings can be bolted to the PC board or heat sink. Heat sinking is quite important, as the internal protection circuitry will shut the regulator down if power dissipation exceeds 7.5W.

13V MOBILE EQUIPMENT

The power transformer (Fig. 2-47) that was pressed into service was a triad F-32A. This particular transformer has two separate 6.3V AC windings, each of which are rated at 3A. These windings were put in series to provide 12.6V AC. It must be remembered that when placing two windings in

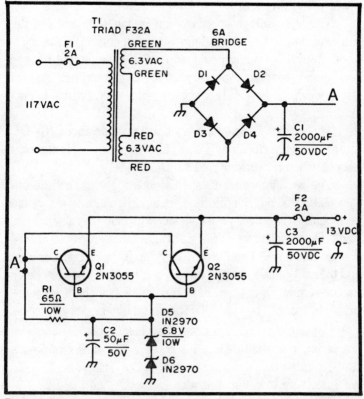

Fig. 2-47. Schematic of the 13V 2A regulated supply for mobile equipment. This circuit is perfect for those mobile units that you retire to indoors, such as transceivers, tapeplayers, scanners, etc.

series, they must be in phase. When you first series the windings measure the resultant AC voltage. If it is not 12V AC, one set of the windings must be reversed.

In the receive mode, most transceivers draw about 100m at 13V DC. This is with the squelch on and no audio present. This resting current will increase with audio; however, the average receive current is low. When in the transmit mode, the current increases to 2 or more at 13V DC. Therefore, it was necessary to design a supply which could provide in excess of the 2A needed in the transmit mode.

Just about any filament transformer with a single 12.6V AC winding or two 6.3V AC windings and adequate current rating (3A) will suffice in this supply. Should nothing be available in the junk box, an old TV set transformer can be used if space is of no concern.

The high voltage windings can be taped up and just the 6.3V AC and 5.0V AC windings placed in series. In older TV sets, the 6.3V winding provided normal tube filment voltage and the 5V winding took care of the 5U4 rectifier. Both of these windings generally were capable of considerable current.

These two windings in series will provide about 16V DC after the rectifier bridge. This is adequate to supply the regulator, which is working at 13V DC.

The rectifier is an encapsulated bridge that can be obtained at Radio Shack outlets. These particular bridges should be mounted directly on the chassis, which acts as a heat sink. Radio Shack sells two different bridges that are rated at 50V 6A and 200V 6A. The 200V 6A bridge is more conservative and tolerant of line surges, and thus might be a better choice for the supply. Of course, individual diodes can be used in the bridge but should be able to withstand the necessary current.

Nothing is more bothering than having components run hot under a normal load. Therefore, a pair of 2N3055 pass transistors were used to regulate the 2A load. The 2N3055 will handle up to 4A so the pair easily handles the 2A.

Although 12.6V AC is put out by the transformer, the bridge rectifier produces approximately 16V to 18V DC. The pair of 1N2970 zener diodes in the base circuit of the 2N3055s holds the output voltage down to 13V DC. The 65Ω, 10W resistor limits the current to the zener diodes so that the dissipation of the diodes is not exceeded.

The reason that two 6.8V zeners were placed in series was that we did not happen to have a 13.6V zener handy. A 1N2979 (15V 10W) might be used but the output voltage might be a bit high...in the neighborhood of 14V DC.

Of course, the use of this supply is not limited to use with a transceiver. Just about any modern transistor-type mobile unit in the 10W power range will work well with this supply.

Parts List

C1, C3.............................2000 μF, 50V DC electrolytic capacitor.
C2.....................................50μF, 50V DC electrolytic capacitor.
D1-D4...................200V, 6A encapsulated bridge (Radio Shack).
D5-D6....................................1N2970, 6.8V, 10W zener diode.
F1..120V AC, 2A fuse.
F2 ....12V DC, 2A fuse.
Q1-Q2......................................2N3055 NPN silicon transistor.
T1 ....Triad F-32A transformer, 115V
AC primary, two 6.3 V AC, 3A secondary windings.

9V TO 28V 10A SUPPLY

The circuit of Fig. 2-48 is a rather standard series voltage regulator in the commercial field but needs a bit of explaining here because it is rather rare in hobbyist usage.

The easiest way to understand how the circuit works is to imagine the series transistor Q1 as a variable resistor that is automatically controlled. If the input voltage E_{IN} were to increase, or the current being drawn by a load at E_{OUT} to decrease, the resistance across Q1 must increase to keep the voltage at the E_{OUT} terminals constant. Similarly, if the load current goes up then the resistance of Q1 should decrease to keep E_{OUT} fixed.

Now we might want to draw quite a bit of current through Q1 maybe as much as 10A. Since the current required by the base of Q1 is approximately the current through the transistor divided by its amplification factor (which may be any where from a value of 10 to about 30) a bit of DC amplification is in order. Q2 and Q3 provide this amplification in a so-called Darlington configuration. Darlington circuits are simply those that hook the base of one transistor directly to the emitter of another, providing a very simple means of increasing gain. In effect, the gain of Q3 is multiplied by the gain of Q2, which is multiplied again by the gain of Q1. Thus a very small current at the input of Q3 can control very large currents through the series transistor.

Now all that remains is to provide some sort of feedback network which will sense the output voltage and control the series transistor. The sensing of the output voltage is done with a simple resistor string across the output, in this case the R6-R7-R8 string. Also, we need a stable voltage source as a reference, and you will note a simple zener diode does the whole trick admirably here.

With the zener providing a good stable reference we can now compare the output voltage to the reference and make the series regulating transistor take up the slack. This is accomplished by a simple differential amplifier, of which the operating theory is adequately covered in most transistor manuals.

If you've made it this far you should have a fair idea of how the gadget works, so now to the easy part, the "makings."

Construction and Operation

The circuit shown in Fig. 2-48 can provide any regulated output voltage from 9V through 28V. The value of E_{IN} is best selected as 25% to 50% greater than the desired output voltage. The input voltage can be easily obtained by a simple bridge rectifier and single capacitive filter, as shown in Fig. 2-49. With Q1 in the circuit, a 10A load current may be drawn continuously. Omitting Q1 and connecting Q2 in its place allows 3A maximum.

Both Q1 and Q2 should be mounted on a suitable heat sink, such as a chassis, with an area of 20 square inches or more. Use the insulating material supplied with the transistors to keep the cases electrically isolated from the chassis, as the collectors are connected internally to the case.

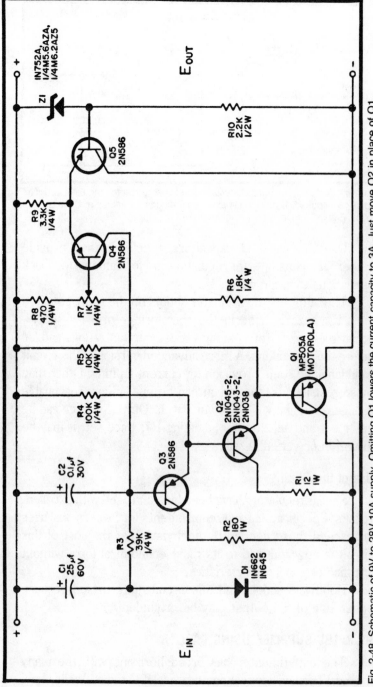

Fig. 2-48. Schematic of 9V to 28V 10A supply. Omitting Q1 lowers the current capacity to 3A. Just move Q2 in place of Q1.

193

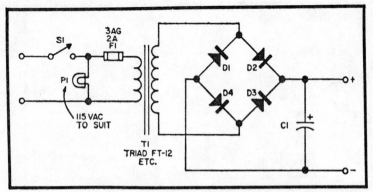

Fig. 2-49. Schematic of unregulated supply suggested for the regulator circuit in Fig. 2-48. Diodes D1 through D4 can be 1A diodes for values of C1 less than 2000 μF and 3A diodes for anything greater.

Use the value of capacitors specified; they provide proper time constants for regulation and for damping feedback oscillations that can occur.

The output of the regulator should not be short-circuited even if a fuze is used in the transformer primary supply, because either Q1 or Q2 may exceed current ratings and fail before the fuze opens. A more advanced version of this circuit would include some provision for current limiting. Other than these restrictions, the circuit can be built in any reasonable configuration and will work quite well. Output voltage can be easily set and adjusted by adjusting R7; once set, it may be forgotten or varied at will.

Use of the Circuit

Similar regulators have been built and used with excellent results. Specifications and requirements for professional uses far exceed those needed by hobbyists, yet the cost of this circuit is now reduced to its most economical form without sacrifice of good characteristics.

Hobbyists should find this circuit quite useful, because either side of the output may be grounded.

9V & 18V SUPPLIES USING CA3018

The experimenter has broad horizons with the many low-cost ICs and semiconductors available today. While many of the ICs have been designed for a specific service, such as an audio amplifier or a logic switch, there are types which contain several transistors to be used in almost any application. The RCA CA3018 exemplifies the second category. Let's examine the device as an introduction to integrated circuitry.

Four silicon transistors are formed on a common monolithic, substrate within the CA3018. Two of these transistors are interconnected by having the emitter of one tied to the base of the other. Either of these transistors may be used separately since an external lead connects to the emitter-base link. The intent of the interconnection is to use the transistors in a Darlington circuit. The other two transistors are isolated as shown in Fig. 2-50. Notice the substrate is important enough to have its own lead. In this IC, as in most others, there is diode action between the substrate and some elements of the circuit. You can see this characteristic using an ohmmeter between the collectors and the substrate. These diodes are reverse biased by connecting the substrate to the most negative point in the circuit, thus isolating the transistors. If the substrate is not connected in such a fashion, you may not get transistor action from the circuit.

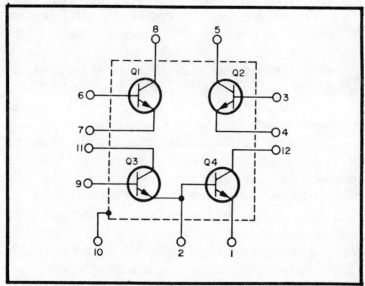

Fig. 2-50. Schematic of RCA CA3018 showing IC pin numbers.

195

Parameter	CA3018	CA3018A
Maximum power dissipation:		
any single transistor	300 mW	300 mW
total package	450 mW	450 mW
Derate 5 mW/°C for TA		
>85°C		
Maximum collector-to-emitter voltage	15V	15V
Maximum collector-to-base voltage	20V	30V
Maximum collector-to-substrate voltage	20V	40V
Maximum emitter-to-base voltage	5V	5V
Maximum collector current	50 mA	50 mA

Fig. 2-51. CA3018 and CA3018A parametric limitations at 77°F.

The transistors in the CA3018 are useful from DC to 120 MHz. One of the big advantages of integrated circuitry is the matched characteristics of the transistors within. Gain, for example, is matched better than 10% and the base-emitter voltage match is better than 2 mV over a wide temperature range. Because of these characteristics, this and other integrated circuits are excellent for temperature-compensated circuitry.

Another plus for most ICs is the excellent low-frequency noise figure. Transistors in the CA3018 array boast 3.2 dB of noise at 1 kHz. At 100 MHz, the noise figure is typically 7 dB, so this device is favored for operation below VHF.

Learning to use the best characteristics of a device and learning to design around its limitations are good engineering practices. Maximum and minimum values for the CA3018, which must be observed, are listed in Fig. 2-51.

Those readers who are familiar with transistors will find only two new ratings; a total-package power rating, and a collector-to-substrate voltage rating. The first is a limit for the sum of the power dissipation of the individual transistors. For example, if one transistor operates at the 300 mW level, the average power of the remaining three may not exceed 50 mW each [300 + (3 × 50) = 450]. The second new rating refers to the breakdown voltage of the collector-to-substrate diodes. These diodes are always reverse biased because the substrate has been connected to the most negative part of the circuit. It is easy to determine the exact voltage in a circuit because the collectors are frequently at the most positive voltage.

Many ICs have low breakdown voltage ratings, which can be a serious disadvantage. In some cases, you may be able to design around the problem. Another problem at the experimenter's level is the fact that if one transistor is destroyed, the whole IC has to be replaced. This disadvantage is somewhat offset by the low cost of the CA3018, but it is hard to rationalize soldering 24 leads for each mishap. A discrete transistor could be used in a crisis with the CA3018. Better yet, use a socket for breadboarding.

Applications

The CA3018 is a natural choice for a power supply regulator amplifier. Such an application is also interesting for analysis. In Fig. 2-52, a very basic regulated power supply is shown using three of the transistors. Let's examine its characteristics from the ratings of the IC.

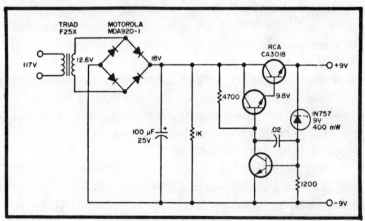

Fig. 2-52. Schematic of 9V supply using the RCA CA3018 as a regulator. Good for 50 mA.

197

Nearly all the current to a load will flow through Q4 (pins 12 and 1); therefore, we are limited to 50 mA, which is the maximum collector current of any single transistor. If you think about it, this is all the current required by many circuits. The power supply could service almost any circuit now operated on small 9V batteries. An FM portable receiver draws peaks of less than 20 mA; AM radios draw even less.

The second rating to consider is the collector-to-emitter voltage. The greatest permissible value is 10.5V, which is well under the 15V rating of the transistor. The collector-base breakdown voltage is not exceeded and the emitter-base junction is never reversed biased, so the design is clean in those respects. Don't forget the collector-to-substrate rating. In this case, the maximum voltage seen is 18V: okay for this circuit, but not for the next one.

When considering the power rating, it is safe to assume we will not exceed 85°C (no derating necessary), and the power dissipation of Q4, (Fig. 2-50) will be much greater than the total power of Q1 and Q3. At 50 mA of current, the power $(P = IE)$ would be $0.050 (18 - 19) = 0.450W$, which is high for a single transistor. A reduction in output current or lowering the input voltage on pin 12 is necessary to operate within ratings. In any case, it would be wise to heat sink the IC for this circuit. You may expect it to get hot above 30 mA.

The shortcomings of the CA3018 become apparent as we attempt to design a more versatile supply such as the one in Fig. 2-53. A 2N5295 power transistor was added to the basic circuitry in order to operate to supply to 18V and 500 mA. Other changes include the use of a temperature-compensated zener diode, variable voltage output, and current limiting by use of the fourth transistor on the IC. Incidentally, you might just want to build a power supply like this if you are newly acquainted with semiconductors.

The 2N5295 will handle 500 mA and higher voltage with ease. If we assume the beta of that transistor is 50 at full load, the base drive required would be 500/50, or 10 mA. That is no problem for Q4 of the CA3018. On the other hand, it soon becomes apparent that we need higher than 20V breakdown.

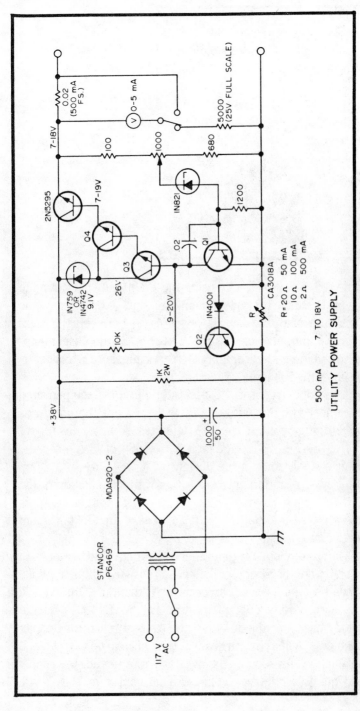

Fig. 2-53. Schematic of 18V supply using the RCA CA3018A along with a 2N5295. Good for up to 500 mA.

199

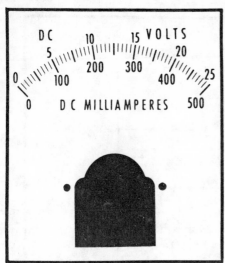
Fig. 2-54. Suggested faceplate for current monitor of the 18V 500 mA version.

Some relief is obtained by using the CA3018A, but not quite enough to handle the collector-emitter voltage drop across the Darlington pair. By using a zener diode between the high-voltage point and pins 11 and 12 of the IC, the collector-to-emitter voltage requirements are reduced, as is the collector-to-substrate voltage.

It is not necessary to use a temperature-compensated zener here. Collector-to-emitter voltages are the only ones above ratings now. Three CA1038s tested had collector-to-emitter breakdown voltage ratings in excess of 23V; however, this is a weak point in the design.

Power requirements are lower with the addition of the 2N5295. The maximum power through Q4 is $23V \times 0.010A = 0.230$ W, enough less than the maximum rating that no heat sink is necessary.

A word about the operation of the current limiter is in order for any prospective builder. Transistor Q2 is normally turned off until an overload occurs. At this time, the voltage drop across R is enough to turn on the 1N4001 and the emitter-base junction of Q2. As the transistor is turned on, its collector is pulled toward ground, thus turning off Q3, Q4, and the 2N5295. As soon as the overload is removed, the supply recovers to its former voltage. This limiter may be made variable for limiting at lower currents. Approximate limiting values are shown in the schematic. Under extreme current limiting conditions, the collector-to-base voltage of Q3 and Q4 will approach 26V. This is acceptable for the CA3018A.

While the CA3018 is not truly a circuit, it provides an interesting introduction to the technology and characteristics of integrated circuitry. Its versatility invites experimentation. Figure 2-54 shows a full-size copy of the meter panel suggested. This can be copied and glued to the face of an appropriate 0-5 milliammeter.

3.5 IC SUPPLY

The circuit shown in Fig. 2-55 was designed to supply an IC logic section of a business machine. It would make a compact low-cost supply for an IC keyer or for general experimental work.

Basically, it takes advantage of the 700 mV forward voltage drop of silicon diodes which varies only slightly with current change. A little time spent selecting individual diodes can provide a regulator for the exact voltage desired, including those values for which standard zeners are not available. Best of all, it can be assembled from parts commonly lying idle, thus freeing the regular bench supply for more demanding work.

The 4.1Ω limiting resistance provides short-circuit protection for the 750 mA rectifiers, and is a necessary part of the filter circuit. A load current variation from 0 to 50 mA causes a

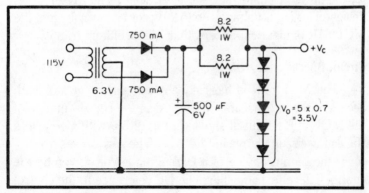

Fig. 2-55. Schematic of simple 3.5V IC supply.

supply drop of just 1.65%. To improve regulation at higher currents, increase the size of the filter capacitor.

5V OR 10V SUPPLY USING MC1460R

The function of a DC voltage regulator is to maintain a ripple-free output voltage that remains constant in value, independent of changes in input voltage, load impedance, and temperature. A complex circuit must be used in order to construct a regulator that provides excellent regulation, fast response time, low output ripple, low output impedance, transient-free output, remote voltage sensing, and current limiting. Also, the number of components necessary makes the constructed circuit occupy a relatively large volume if discrete components are used. However, a complex regulator exhibiting good regulation can be constructed as an inexpensive integrated circuit and assembled into a package that occupies a very small space. The Motorola MC1460R monolithic voltage regulator is such a circuit; it is packaged in a single 9-pin TO-66 case, yet can supply an output current of 500 mA with no external semiconductors. This device can reduce the parts count and the volume required for a regulator while maintaining excellent operating characteristics and low cost.

The MC1460R can be combined with a small etched circuit board to form a complete, well regulated power supply that can be built in a space not much larger than that required for the power transformer and filter alone. The MC1460R is designed for input voltages up to 20V; for higher voltages, the MC1461R is identical except that it is useful up to 35V.

Circuit Operation

The MC1460R is basically a series-pass regulator with access to the internal amplifiers. A discrete representation of the MC1460R circuit is shown in Fig. 2-56, with a complete regulating circuit shown in Fig. 2-57. The series-pass transistor is the output device of a Darlington circuit driven by the control differential amplifier. The DC reference to the control differential amplifier is provided by an internal series reg-

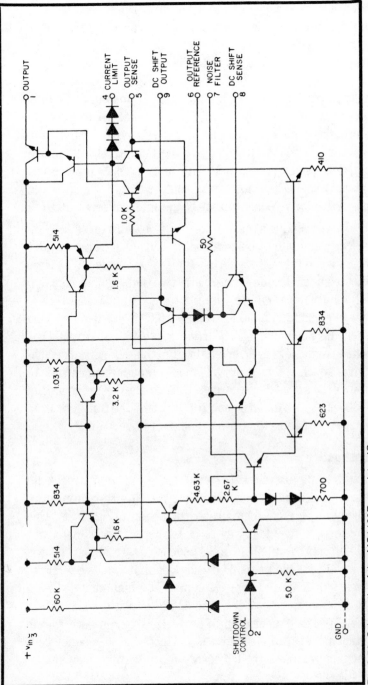

Fig. 2-56. Schematic of the MC1460R regulator IC.

203

ulator. The reference input to the internal regulator is derived from a zener which is buffered by a transistor amplifier. This voltage is divided by a series resistor-diode string. The output voltage of the internal regulator, which appears at pin 9, is determined by resistors R2 and R3 (Fig. 2-57), connected between pins 9 and 8, and pin 8 and ground. Pin 9 is then connected to pin 6 to provide a reference to the control differential amplifier. The other input to the control differential amplifier (pin 5) is tied directly to the output. Thus the main control amplifier is operated with unity gain to provide optimum AC and DC performance, and the output voltage of the main regulator is equal to that of the internal regulator.

By connecting a resistor from pin 1 to pin 4 (in series with the load), the output current can be sensed and limited. This occurs when the voltage across R4 is large enough to forward bias the internal diode string. Since two of the diodes compensate for the base-emitter drops of the output transistors, a voltage equal to one diode drop is sufficient for this to occur. When this happens, base current drive to the output Darlington transistors is diverted through the diodes. An equilibrium is reached which provides just enough base current to forward bias the diode string. Due to the large amount of feedback used in the main control amplifier, the transition from voltage regulation to current limitation is very sharp.

Construction

All of the parts for the regulator in Fig. 2-57 except the power transformer, bridge rectifier, and filter capacitor can be placed on the etched circuit board shown in Fig. 2-58. This provides a convenient way to construct an operating circuit and also controls the lead dress used in construction. This will minimize possible parasitic oscillations that may occur due to phase shift from reactive elements. The high gain and wide frequency response of the monolithic regulator makes this a problem at frequencies up to 60 MHz, so even stray wiring capacitance can cause oscillation. Thus, high frequency construction practices are necessary to prevent oscillation possibilities.

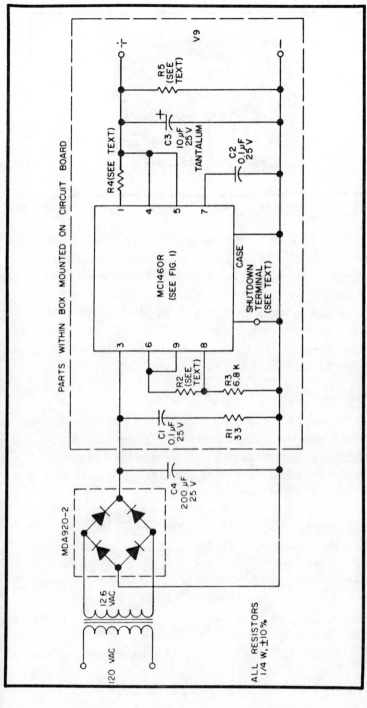

Fig. 2-57. Regulated power supply using the MCi460R/MC1460R IC.

205

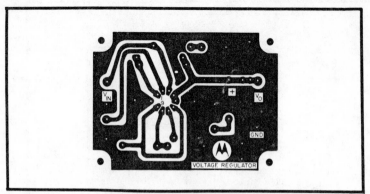

Fig. 2-58. Suggested PC board for the MC1460R and its associated circuitry.

The MC1460R must be mounted on a heat sink for high power dissipation. In this case, the passive components are mounted on the copper side of the board. The heat sink must be spaced away from the board to prevent shorting to the leads of the components which project through the board. Silicone grease should be used between the IC and the heat sink for best heat dissipation, and insulation should be used on all leads to prevent shorting to the heat sink.

Output Regulation

The curves of Fig. 2-59 show that excellent voltage or current regulation can be obtained with the assembled circuit. No heat sink was used for these curves. For the regulator used, the output voltage varied only 2 mV for the 5V setting from 1 to 120 mA of load current, which corresponds to a load regulation of 0.04%. Less than 10 mV of change in the output voltage could be observed for the 10V output with up to 130 mA output current, which is less than 0.1% load regulation. This performance results from the combined effects of output impedance and junction heating. The effect of output impedance tends to make the output voltage fall while junction heating effects from increasing load current tend to make the output voltage rise.

The high gain of the MC1460R makes the transition region between the voltage, and current-regulating modes very small and the current-limiting region very sharp. Thus, from 9.99V to 0V output, the current varies by only 3 mA from its limiting value of 150 mA.

Input Regulation

The DC input regulation characteristics are shown in Fig. 2-60 for load currents of 1, 50, and 100 mA for both 10V and 5V output. Again junction heating effects are present and result in the 0.01% per volt regulation shown. The AC input regulation, although not shown, is typically 0.003% per volt.

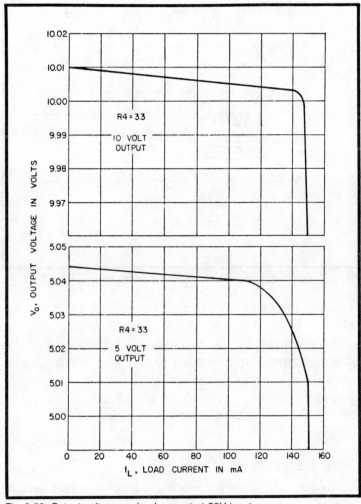

Fig. 2-59. Output voltage vs. load current at 20V input.

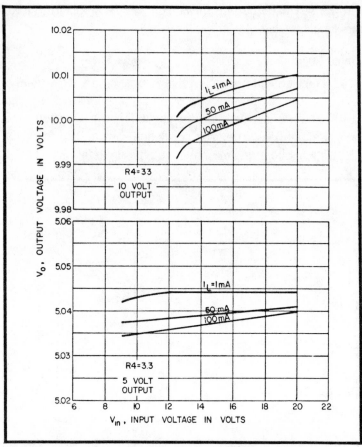

Fig. 2-60. Output voltage vs. input voltage at various load currents.

The minimum input-output voltage differential for the MC1460 is 3V. Thus, when the output is set for 10V, the required input is 10V plus 3.0V, or 13.0V. For 5V output, the minimum input-output voltage requirement would give an input voltage of 8.0V. However, the minimum input voltage for proper operation of the MC1460R is 8.5V, and this would be the required input voltage for anything less than 5.5V output.

Power Dissipation Limitations

When the input voltage is at the maximum rated input of 20V and the output is shorted, the regulator must withstand

208

this voltage and supply the short-circuit current Isc, which is itself set by R4 according to the following formula:

$$R4 = \frac{0.5V}{\text{max output current}}$$

For example, with an R4 of 3.3Ω, Isc will be 150 mA. For 20V in, 0V out, this gives a power dissipation of 3W. Therefore, the package would have to be mounted on a heat sink. The amount of heat sink area required depends on the ambient temperature and in this example would have to be enough to keep the case below a temperature of 127°C.

Output Voltage Adjustment

The output voltage is set by the values of resistors R2 and R3. The voltage at pin 8 is typically 3.5V. To minimize the loading effect due to base current on the voltage at pin 8, a current of about 0.5 mA is required through resistor R3; thus, its value should be about 6.8K. Although a low temperature coefficient is not required for these resistors, the *same* temperature coefficient is necessary to keep the ratio independent of temperature so that the output voltage is constant over temperature.

Shutdown Control

Both the MC1460R and the load current can be turned off (or shut down) via the shutdown control at pin 2. Under this condition the output voltage drops to almost zero, allowing only a few micro-amperes to pass through the load. The shutdown control voltage can be provided by any saturated logic such as DTL, RTL, TTL, etc. If shutdown is not necessary, pin 2 should be grounded to prevent spurious signals and inadvertent connections from shutting down the regulator.

Conclusion

The basic MC460R integrated circuit voltage regulator is seen to be a useful component which operates with a minimum of external components and without the need for external semiconductors. The outstanding performance achieved by the regulator can be obtained at a relatively low cost.

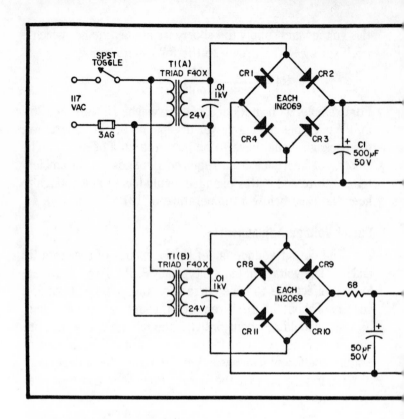

ZERO TO 30V LAB SUPPLY

A typical laboratory power-supply has variable voltage output, low internal impedance, good voltage regulation with a variety of loads, freedom from output changes with line-voltage fluctuations, adjustable current limiting, low ripple and noise voltage in the output, and accurate metering of output voltage and current.

The laboratory power-supply circuit is shown in Fig. 2-61.

Power transformer T1 is two Triad F40X's, both operated in the full-wave bridge configuration. The series transistor (Q1) is listed as a 2N375, but a number of other 80V power transistors have been used in its stead, including the 2N174 and 2N1542.

Most of the circuitry is laid out in a simple etched circuit board as shown in Figs. 2-62 and 2-63 and so wiring is very

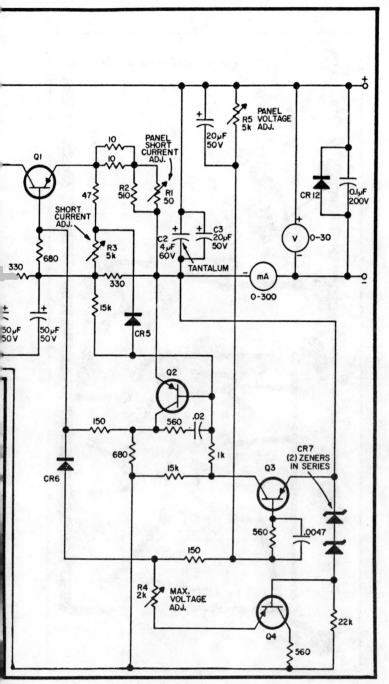

Fig. 2-61. Schematic of 0 to 30V lab supply.

211

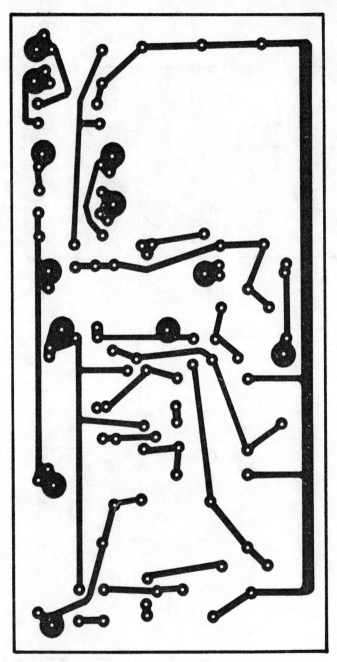

Fig. 2-62. PC board diagram for the 0 to 30V lab supply.

fast. The supply is housed in a standard LMB-W1A cabinet. The series regulator (power) transistor is heat sink mounted on the back plate (with a mica insulator, of course).

A silicon rectifier was added across the output of the supply in addition to the 0.1 μF capacitor to help kill any transients (from the equipment being run by the supply) that are opposite to the supply polarity.

The meters used are the inexpensive Japanese-made miniature types available under many U.S. names including Calrad, Monarch, and Lafayette. Their accuracy seems adequate for this application.

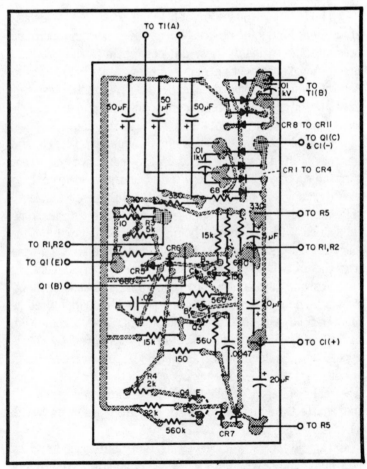

Fig. 2-63. Component layout diagram for the 0 to 30V lab supply.

One will note that C2 and C3 are in parallel. The original purpose of this apparent duplication was to assure that the 20 μF electrolytic was a low impedance for low frequencies, and the 4 μF tantalytic was a low impedance for higher frequencies. However, if one has a 25 μF to 50 μF tantalytic, the single unit will do by itself.

When the power supply is finished, just plug it in, adjust the two screwdriver adjustments "max. voltage adj." and "short circuit current adjust, for the 0-30V range and 25 to 225 mA range of the front-panel controls.

ZERO TO 30V 60W SUPPLY

The hobbyist today cannot help but notice the wide variety of semiconductor devices being advertised in electronic orientated magazines. These devices, selling at a fraction of their original cost, were unheard of a few years back. For the hobbyist who likes to build and experiment with these fascinating items, an adjustable, regulated, solid-state power supply is described here that is both easy to build and use.

Basically the unit consists of two supplies, T1, T2, and their associated circuitry make up the main, heavy current carrying supply. T3 and its associated circuitry serves only to provide amplifier chain Q1, Q2, and Q3 with a well-regulated reference voltage. Both supplies are common series type regulators, with Q1 and Q4 being the regulating element for each. Amplifiers Q2 and Q3 are included to isolate Q1's base from the 5K pot, so that only a small current is needed at Q2's base to swing the large base current at Q1. A close look at Fig. 2-64 will show that all the transistors are being operated as emitter followers, or current amplifiers. The output from diode bridge D5 through D8, after being filtered by C2, is applied to the collector of Q4 and the 620Ω resistor. The 620Ω resistor is picked to fire zener D3 with about 10 mA of current. The zener voltage is applied to the base of Q4 and appears at the emitter, minus the 0.6V or so drop associated with silicon-type devices.

The 5K linear pot serves as the emitter load and by tapping up and down this resistor, a voltage of from 0 to 33V

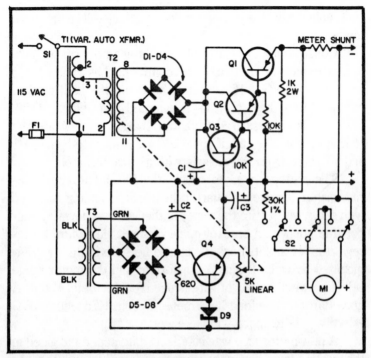

Fig. 2-64. Schematic of 0 to 30V 60W supply.

approximately is available to control amplifiers Q3 and Q2. Autotransformer T1 is coupled mechanically to the shaft of the 5K pot. This is done so that no matter what the voltage of the unit is set at, the drop across Q1's collector to emitter is constantly 4V or so. If this were not done, using the unit at say 9V would require the series element Q1 to dissipate the voltage difference between 9V and the total voltage that would appear across C1. The voltage across C1 is about 40V without the autotransformer so we have 40V minus 9V, or 31V times the rated current of 2A or a dissipation of 62W. Since the output rating of the supply is 60W (30 × 2) we see that more wattage is being used to heat the room than is available for our bench projects. Putting T1 in ahead of T2 modifies the voltage available at C1 so that approximately 5V more is available at C1 than is supplied to its base by the reference supply. So the dissipation of Q1 is 5 × 2 or 10W, and this is just lukewarm. Since most transistor projects require considerably less current, unless you are power minded, the unit will run at room temperature.

Construction and parts placement are not critical. Two items to watch for are that the 5K control pot has a linear taper and to use AWG #16 or better wire for the high current carrying half of the supply. Q1 is mounted on a Delta heat sink NC-401 that is insulated from the chassis, or insulate the transistor from the sink if you want. Q2 is mounted under C1 and insulated from the chassis with a mica insulator.

The method used to couple the 5K pot to the autotransformer involves a quarter-inch shaft coupling fastened with epoxy cement to the wiper of the autotransformer. Support the pot with a bracket made of sheet aluminum. Series regulator Q4 and its socket can also be seen in the top view mounted between the autotransformer and the meter range selector switch. A coil of wire is used for the meter shunt. A good starting place for a 1 mA meter is about 1.75 feet of AWG #16 enameled copper wire.

A new meter face was inked on white paper and glued to the existing face. The new ranges are 0 to 30V and 0 to 3A.

To adjust the pot and transformer, set them initially at their maximum voltage positions. Do this without the AC plug in. Lock the pot shaft to the autotransformer. Bring the pot and transformer back to zero and plug the supply into the AC outlet. Put a voltmeter across the collector to emitter of Q1 and turn the switch on. Set the supply for about 10V output and note the reading across Q1. If it is anywhere from 4V to 6V, you are in business; if not, turn the switch off and not disturbing the pot or transformer unlock the shaft. If the voltage drop was higher than 6V, hold the pot shaft from turning and move the wiper of the transformer about a quarter-inch back toward its zero position and lock the shaft. If it was lower, move the wiper up about a quarter-inch toward its maximum voltage position and lock the shaft. Turn the unit on and again measure across Q1. Continue to do this until the correct voltage drop is achieved.

Parts List

T1..Superior Powerstat type 10B
T2..Stancor type RT-202

216

T3 ....Stancor type P-6469
D5, D6, D7, D8................................100 PIV 100 mA diodes
            or better as: Texas Inst. 1N537 or Motorola 1N4002
D1, D2, D3, D4..3A to 5A 100 PIV diodes or better: GE 1N3569
Q1, Q2...Motorola 2N376A
Q3, Q4 ....Motorola 2N1132
D9 ....33V 1W zener diode Sarkes Tarrian VR-33
C1 ....1500 μF 50V capacitor Sprague TVL-1341
C2, C3......................100 μF 50V capacitor Sprague TVA-1310
M1 ....1 mA meter
S1 ....SPST switch
S2............................3 pole 2 position rotary switch
F1..1A fuse

ZERO TO 24V SUPPLY

How often have you picked up some goodies at the local suplus market and rushed home to hook it up for trial, only to find that your 12V car battery or battery eliminator did not supply enough of the 24V needed to make it go? Or when the kids' transistor radio quit working, did you run all the way down to the radio store and buy a new battery only to find that it wasn't the right battery after all? The variable power supply described here can help you out of these situations as well as many others. You can even toss the battery eliminator out and use this unit to charge the car battery, or convert the eliminator into the type of supply described here, using many of the original components.

This supply is simple to construct, small in size, economical to build (if you have the major components in your junk box), has very smooth control, good regulation, and excellent filtering.

Many of the components used were not labeled as to part number, manufacturer, etc., but their ratings are similar to those shown. This same circuit can be used for supplies other than 0 to 24V, just don't exceed the voltage and current ratings for the diodes and transistors used.

About the Circuit

T1 as shown on the schematic in Fig. 2-65 is hooked up to a conventional bridge rectifier circuit. If you happen to have a center-tapped secondary transformer and only a couple of diodes, the circuit will work just as well with that arrangement. Likewise, a higher voltage transformer, up to about 50V, would make the supply even more useful (a 36V 5A transformer was used here).

The first transistor, Q1, is the filter regulator part of the circuit. It is hooked up in what is sometimes called a capacitance multiplier circuit. That is, the filtering at the points marked A and B on the schematic is equal to the beta of Q1 multiplied times the capacitance of C1. For instance, if the beta of Q1 is 40 and the capacitance of C1 is 2000 μF the DC output is filtered to the equivalent of an 80,000 μF capacitor across the output. This part of the circuit can also be duplicated to give still additional filtering. If you have no 2000 μF capacitor, but do have a couple of 500 μF capacitors, rather than parallel them into a single unit, use one in the base of Q1, add another transistor and use the other capacitor in its base circuit. This will give you the equivalent filtering of 500 μF times the beta of Q1 (40) or 20,000 μF. The additional transistor will give you an additional 20,000 μF equivalent, and will also improve the regulation somewhat.

The second two transistors, Q2 and Q3, are the voltage control transistors. Two transistors are hooked in parallel to allow for the power that must be dissipated in that part of the circuit and will change with the change of voltage control R4. If necessary a third or fourth transistor can be added in parallel with Q2 and Q3 if your transformer is capable of higher current output than 5A. In that case additional transistors will have to be parallel with Q1 also. On the other hand, if your transformer is only capable of a couple of amps., a single 2N1970 will suffice. Even a cheaper transistor such as the 2N307A may be used for Q1, Q2, and Q3 for lower voltage, lower current requirements. In any case, all three transistors should be mounted on a good heat sink as they will run warm. You don't have to be fussy about components to build yourself a good well filtered power supply. Take inventory of your junk box, decide what voltages and current you would like to have, and if you have the transformer to fill this requirement you can

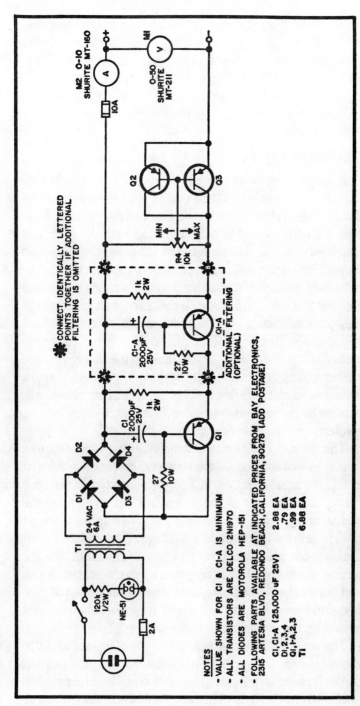

NOTES
- VALUE SHOWN FOR C1 & C1-A IS MINIMUM
- ALL TRANSISTORS ARE DELCO 2N1970
- ALL DIODES ARE MOTOROLA HEP-151
- FOLLOWING PARTS AVAILABLE AT INDICATED PRICES FROM BAY ELECTRONICS,
  2315 ARTESIA BLVD, REDONDO BEACH, CALIFORNIA, 90278 (ADD POSTAGE)

C1, C1-A (25,000 uF 25V)	2.88 EA
D1, 2, 3, 4	.79 EA
Q1, 1-A, 2, 3	.99 EA
T1	6.88 EA

* CONNECT IDENTICALLY LETTERED
  POINTS TOGETHER IF ADDITIONAL
  FILTERING IS OMITTED

Fig. 2-65. Schematic of the 0 to 24V variable supply.

219

use almost any type of diodes, transistors, and capacitors by using them in series, parallel, etc., to fill the current and voltage requirements. Uses for this supply are limited only by the imagination. The filtering is sufficient to allow its use for transistor radios, tape players, etc., yet it has enough current available to charge car batteries or run solid state mobile rigs on the workshop bench.

0.35V TO 15V SUPPLY

A block diagram of the power supply is shown in Fig. 2-66. The unregulated voltage (V_u) is applied to an emitter-follower stage. This transistor acts as variable resistor R_s in series with the power supply and load resistance R_L and maintains the output voltage at (approximately) the voltage on its base. The output voltage is fed back to the summing point. A reference voltage is also applied to the summing point and compared to the output voltage. If the output voltage is not the same as the reference voltage, a difference, or error voltage appears at the input of the amplifier. The error voltage is amplified and applied to the base of the emitter-follower stage with such polarity as to cause the output voltage to approach a level equal to the reference-voltage level. This should immediately be recognized as a negative-feedback amplifier with the reference source acting as the input signal.

The summing point may be treated either as a voltage or as a current summation. Unfortunately, current summing at the amplifier input requires the reference and output voltages be of opposite polarity so that in equilibrium, the sum of the currents into (or out of) the amplifier-input terminal equals zero. There are some definite advantages to be gained by this technique, but a second power supply is required. Voltage summing is employed as it requires no extra supply. As an added advantage, the voltage-summing circuit employed provides some gain for the error voltage in addition to that provided by the amplifier.

The amplifier is necessarily a DC amplifier, and as such, is subject to drift. This condition is made worse by the fact that high gain is necessary for good regulation of the output voltage as load current changes. The amplifier is stabilized by large

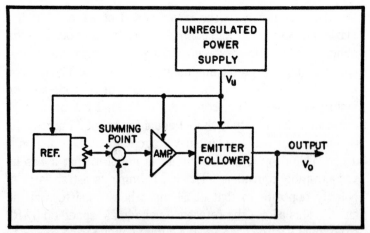

Fig. 2-66. Block diagram of 0.35V to 15V supply.

amounts of feedback. Simply stated, although the amplifier has a large gain, the ratio of the output-to-input voltage is one. The only point not stabilized by the DC feedback is at the input of the summing point, and this has been stabilized by an additional compensating circuit. AC feedback is employed to maintain stability and regulation with capacitive loads and during load transients.

The circuit diagram appears in Fig. 2-67. An unregulated power supply is formed by T1, CR1 through CR4, and C1. The rectifiers form a bridge circuit to full-wave rectify the transformer secondary voltage. With this circuit full advantage is taken of the transformer secondary and no center tap is required. The value of C1 is large enough to remove most of the ripple. It will be noted that its value is smaller than that normally associated with low-voltage/high-current supplies. Remember that the output voltage is tied closely to the value of the *reference*. Thus, if the reference-voltage ripple is low, the output ripple will also be low. If the power supply is poorly filtered and contains appreciable ripple but the regulator output does not, the regulator has *electronically* removed a great deal of the ripple, just as if a huge capacitor had been employed for filtering. C1 may be thought to have been electronically multiplied by the regulator circuit, in this case to a minimum value of about 50,000 μF.

The reference source is composed of R1 and Z1. If a maximum degree of regulation is desired, with voltage summing, the reference voltage must be as large as the highest output voltage desired (remember the 1:1 ratio). The use of a zener diode holds the voltage at the base of Q1 constant. An emitter-follower stage, consisting of Q1 and R2 maintains a constant-voltage output almost equal to that of Z1 across the potentiometer R2, the output-voltage control. Summation is accomplished by Q2. This circuit is novel in that it has no vacuum-tube equivalent. That is, it employs a transistor of polarity opposite to that of all the other transistors in the circuit. This allows the reference voltage as tapped off by R2 to be applied to the base and the output voltage to be applied to the emitter. Any difference between the two voltages is an error signal which is amplified in the collector circuit and applied to the base of Q3, a grounded-emitter circuit. The signal developed across collector load resistor R6 is directly coupled to the base of Q4.

The inclusion of R7 assures the ability to adjust the output voltage to the lowest possible minimum value. The actual regulation is the function of Q5. A minimum load, or "bleed" is provided by R8. Drift of the output voltage, which would be caused by changes in temperature of Q2 by altering the base-to-emitter voltage drop, are compensated for by CR5 and R3. Diode CR5 is forward biased by R3, but its small forward drop is in opposition to the V_{be} of Q2. As the temperature of CR5 and Q2 is raised, both forward drops increase by approximately the same amount, and the voltage between the output and reference remains constant. Returning R3, R4, and R7 to a point slightly more positive than the positive output terminal also helps reduce the lowest minimum output voltage obtainable. The use of a silicon rectifier at CR6 operated in the forward-biased condition provides a small but constant voltage for this purpose. In a sense, it is an economical low-voltage zener diode.

The noise on the output is reduced by the additional filtering of the reference voltage by C2. As with any feedback amplifier, instability may be a problem. The combination of R5

222

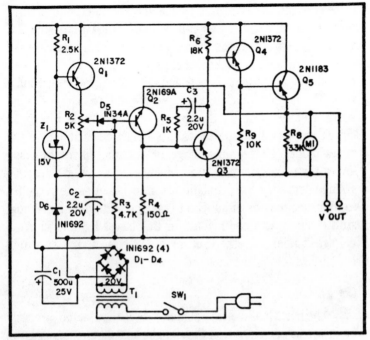

Fig. 2-67. Schematic of 0.35V to 15V supply.

and C3 stabilize the loop by providing degeneration which increases with frequency, necessary when the supply is used with reactive loads.

Operation

With the components specified this supply will deliver 100 mA continuously from its minimum output voltage of 0.35V, to its maximum voltage of 15V. Figure 2-68 illustrates the output-voltage current curves obtained at various values of Vo. The design of this circuit allows considerable latitude in the choice of components. Transistor Q5 may be a higher or lower powered type, though in any case care must be taken that the supply is not operated in such a manner as to exceed the collector-temperature limitations due to excessive current at low output voltage. Proper heat-sinking is absolutely essential. The choice of types for the other transistors is dictated primarily by the voltage supplied from the unregulated power supply (and proper polarity, of course). It is well to be on the safe side and select types which have collector-to-emitter voltage ratings which are approximately 1.4 times the AC voltage appearing across the secondary of T1. Transistors of higher beta should make little improvement in regulation, though lower beta units for Q2 and Q3 will degrade the regulation. The detectable changes of Vo are due to changes in the zener voltage. This is due partly to the simple reference circuit employed and partly to the dynamic resistance of the zener diode. The zener voltage of W1 may have any value lower than that given though the use of a lower reference voltage degrades the regulation somewhat. A diode in the 5V to 6V range may be employed to take advantage of the almost zero temperature coefficient. In this case, R8 must form a voltage divider with the emitter of Q2 returned as shown in Fig. 2-69.

Construction

The parts listed on the schematic were selected primarily on the basis of size so that they might be "shoehorned"into an available box. As a consequence, several special parts were made, such as the heat sink for the power transistor. No details of the etched board are shown as each individual will undoubtedly make some parts substitutions. Some of the obvious ones include the use of a heavier power transformer

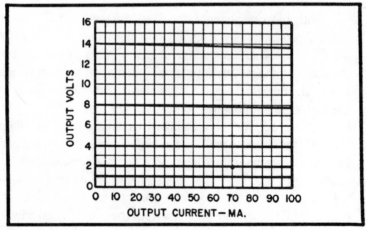

Fig. 2-68. Performance curve for the 0.35V to 15V supply.

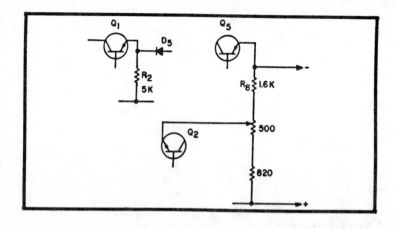

and filter capacitor, and a transistor at Q5 capable of dissipating more power. Remember that adequate heat-sinking is of prime importance. No other form of over-load protection is provided. If higher output voltage is desired, the technique illustrated in Fig. 2-69 may be employed, but the secondary voltage of T1 must also be increased. In the event that circuit voltages are changed (and with some transistor substitutions), it will be necessary to alter some of the resistance values.

Parts List

C1..500 μF, 25V
C2, C3..2.2 μF, 20V
CR1, CR2, CR3, CR4, CR6 ....1N1692
CR5..1N191
R1...2.5K
R2..5K linear taper pot
R3...4.7K
R4 ....150Ω
R5 ....1K
R6...18K
R7...10K
R8...3.3K
SW1...SPST slide switch
T1 ....Power transformer, input 110V, 60Hz Output 20V, 140 mA
Z1 ....Zener diode, 13V
Q1, Q3, Q4..2N1372
Q2..2N169A
Q5..2N1183
M1 ....0-15V DC meter (Lafayette TM-100)

225

Chapter 3
Power Supply Miscellanea

In this chapter there is a collection of several power supply related circuits that will be useful to the hobbyist. Most of the circuits are add-on things that either assist or complement an existing power supply. Some of the add-on types include a voltage splitter, a voltage divider, an IC protection circuit, and a current limiter. Others, more unique types, are such things as a capacitor rejuvenator-leakage tester and a memory holding circuit for digital clocks.

VOLTAGE SPLITTER

While the dual supply approach is not a bad one, the method herein described has several advantages. By plugging this splitter into an existing regulated power supply, a center-tap is electronically created which will stay midway, voltage wise, between the plus and minus terminals. This centertap will stay at the midpoint even if current is drawn from it, and will stay at the midpoint even when one varies the supply voltage. This relaxes the requirement of extremely good supply regulation if the positive and negative voltages change equally.

The splitter (Fig. 3-1) is built on an etched circuit board with a dual male banana-pin plug mounted on it for direct

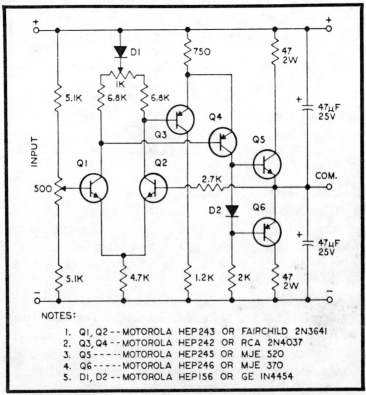

NOTES:
1. Q1, Q2 -- MOTOROLA HEP 243 OR FAIRCHILD 2N3641
2. Q3, Q4 -- MOTOROLA HEP 242 OR RCA 2N4037
3. Q5 ----- MOTOROLA HEP 245 OR MJE 520
4. Q6 ----- MOTOROLA HEP 246 OR MJE 370
5. D1, D2 -- MOTOROLA HEP 156 OR GE 1N4454

Fig. 3-1. Schematic of voltage splitter.

connection to one's regulated supply. Three binding posts are similarly mounted on the board and serve as the output terminals.

By use of economy transistors, the price of parts of the splitter herein described is low. Numbers are given for each semiconductor in the parts list. Motorola HEP numbers are the more commonly available types. The other numbers are perhaps more readily obtainable by those who work in the electronic industries and regularly deal with large industrial electronic supply houses.

The etched circuit board and the parts placement on it are shown in Figs. 3-2 and 3-3. Note that the MJE370, MJE 520, HEP S5000, and HEP S5006 are in what Motorola calls a "Thermopad package." Heat dissipators may be easily and inexpensively made for these transistors out of small

227

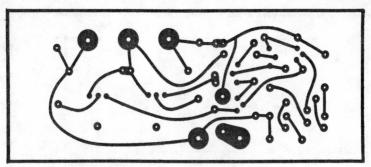

Fig. 3-2. PC board of voltage splitter.

U-shaped pieces of aluminum or copper sheet. The circuit board was laid out to fit Beckman 78PR type pots.

Once the parts have all been soldered to the board you're ready for testing. It's best to try the splitter on a current-limiting type supply, with the current limiting set to about 50 mA.

The regulated supply feeding the splitter is set to 24V, so that the output of the splitter will be ±12V. The reason for calibration at this voltage is that it is in the middle of the useful operation range (10V to 40V input) and also it fits the 12V requirement for operational amplifiers. The splitter will draw about 15 mA with no load; if it draws considerably more, check for soldering errors or faulty semiconductors. The 500Ω pot is adjusted until the plus and minus outputs are equal. The setting of the 1000Ω pot is more tedious; the setup is shown in Fig. 3-4. Offset of the centertap from a true split is indicated by the VTVM as the input voltage is varied form 10V to 40V. The offset is corrected by adjusting the 1K pot, and again sweeping the input voltage from 10V to 40V. This procedure is repeated until the best approximation to perfect splitting is obtained.

The splitter can deliver up to about 100 mA through the centertap, but this should rarely be necessary since the main current drawn by operational amplifiers and other linear ICs is between the plus and minus terminals. The splitter has been used to power Fairchild μL709s and National LM201s on ±15V, General Electric PA223s on ±12V, and General Electric PA238s and RCA CA3029s on ±6V, all with success. As

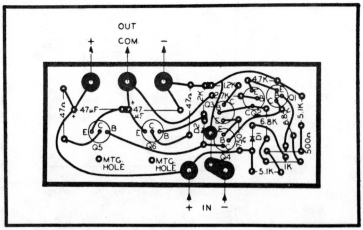

Fig. 3-3. Parts layout of voltage splitter.

the world of linear ICs expands, this simple power supply splitter may easily become the handiest piece of bench equipment you own (next to the basic regulated power supply, of course).

Parts List

Circuit board..A.R.S. Enterprises, P.O. Box 555, Tempe AZ 85281
Input plug.....................................Pomona MDP-ST sawed off
Output terminals.....................................General Radio 938-C,
938-D, 938-G (one each) and three 938-F
Q1, Q2.....................Motorola HEP S5014 or Fairchild 2N3641
Q3, Q4.........................Motorola HEP S5013 or RCA 2N4037
Q5.....................................Motorola HEP S5000 or MJE 520
Q6.....................................Motorola HEP S5006 or MJE 370
D1, D2...........................Motorola HEP R0052 or GE 1N4454

ACTIVE VOLTAGE DIVIDER

If you intend to perform experiments that require dual power supplies and you only have a single-ended supply, then the circuit in Fig. 3-5 is for you. Pin numbers are for the mini-Dip or TO-5 packages.

Actually, it is nothing more than a voltage follower, just textbook stuff, but maybe not widely known in hobbyist circles. As its name implies, the operational amplifier, by virtue of its intrinsic properties, will cause the voltage at its output to

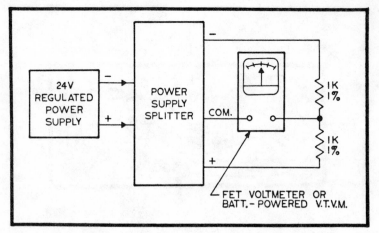

Fig. 3-4 Test setup for adjustment of the pots on the splitter.

follow the voltage impressed upon its noninverting input. This voltage can be conveniently selected by means of a potentiometer.

The op amp's output will then play the role of a "synthetic" ground, and you will enjoy the possibility of selecting the potential of the ground terminal, provided that it does not come closer than 3V to either supply line.

In addition to the above limitation, the maximum current output of the op amp will dictate the maximum current differential between the two branches of the circuit that it can

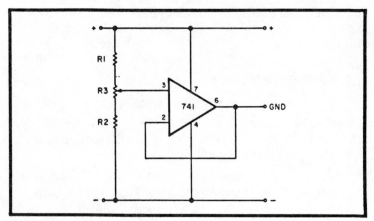

Fig. 3-5. Schematic of basic voltage divider. Pin numbers are for the mini-DIP and TO-5 packages.

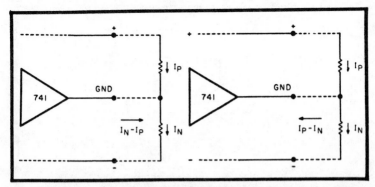

Fig. 3-6. Direction of current flow through the 741. (Top) Current in the positive branch is greater (Bottom) Current in the negative branch is greater.

handle. If the two currents are identical, the op amp will be just loafing along and, theoretically, there will be no current whatsoever flowing into or out of the op amp. If, however, one of the two branches draws more current than the other, the op amp will have to source or sink that difference. See Fig. 3-6. The arrows indicate conventional current flow.

The 741, like most other op amps, is internally protected against short circuits and can indefinitely withstand a short between its output and either supply line. Therefore, do not be afraid of blowing up the device. If you try to exceed its capabilities, it will just refuse to cooperate, and you will lose control of the voltage at the ground terminal.

This limitation raises another question: how much current can you expect out of this circuit? There is not one single answer to this question. If the positive and negative currents are nearly identical, the limitation will be imposed by the capabilities of the single-ended supply, but if, on the other hand, those currents differ widely, the limitation will be imposed by the maximum output of the 741. It so happens that this device was never intended to be a current driver and, in fact, manufacturers do not guarantee or even specify the current that you can expect out of one of those little beasts. If you sift through available literature, you will see this parameter quoted at anywhere between 5 and 15 mA.

Confronted with this situation, it was decided to find out. Some twenty-five different units from several manufacturers

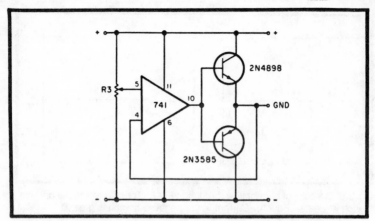

Fig. 3-7. Schematic of a modified version of the basic active voltage divider. This version will handle heavier currents. Pin numbers shown are for the 14-pin DIP.

were gathered in different types of packages. Their short-circuit current was measured with unexpected results: mini-DIPs and metal cans gave about 16 mA, whereas 14-pin DIPs gave a surprising 30 mA.

Inasmuch as the output current limitation is temperature dependent, some tests were run with heat sinked DIP packages with inconclusive results, and it is doubtful whether it is worth the trouble.

There are in existence other op amps capable of greater current output, but they are not only difficult to find, but expensive, as well. The circuit of Fig. 3-7 is a more practical solution if the unaided 741 is not capable of delivering the performance you need. This circuit was designed after unsuccessfully trying two different designs. The first design only worked if the negative branch drew a heavier current, and the second one was plagued by oscillations.

In the unit shown in Fig. 3-7, 2N4898 and 2N3585 transistors were used for the simple reason that they were available. Any power transistor in a TO-66 or TO-220 package should be okay. It would be futile to try a big brute like a 2N3055 because the 741 would be unable to provide the base current necessary to fully exploit its power capabilities. With the transistors above, the unit can provide currents with an imbalance exceeding 1A.

232

If you are still not satisfied with that, you may use Darlington amplifiers instead of plain transistors. These amplifiers from the outside look just like any transistor, and they are inserted in a circuit as if they were just that. The main characteristic of such devices is a phenomenal static current gain. Therefore, the skimpy current output of even a 741 will be sufficient to provide the necessary base drive. Suitable devices would be Motorola's HEP S9101 and HEP S9121. Texas makes the TIP 110 and the TIP 115, which are lower in power capability and would require heat sinking. See Fig. 3-8 for the main parameters.

On the schematic, R1 and R2 are optional and, if installed, should be about 10% of R3. The pot should be about 50 K and the bypass capacitors (from ground to the supply rails) should be a 0.1 μF ceramic disc. It will be advantageous to use the 14-pin package, rather than the mini-DIP or metal can, and have all pins soldered to the board rather than socketed, for improved heat transfer.

Before you decide to go the Darlington way, make sure that your power supply has enough beef to exploit the performance afforded by the very high current gain of these devices. Remember that the tremendous amplification is obtained only under static conditions and falls off quite rapidly with increasing frequency. Eventually you may run into a circuit whose current demands fluctuate too rapidly for the Darlington to follow.

INSTANT REGULATION

Need an additional regulated supply in a hurry, perhaps to finish up some project which requires a special voltage, or perhaps the regular supply is just overloaded? If you have a

Device	Polarity	BV_{CBO}	BV_{CEO}	BV_{EBO}	I_c	P_d
S9101	NPN	60	60	5	4 A	40 W
S9121	PNP	60	60	5	4 A	40 W
TIP 110	NPN	60	60	5	2 A	50 W
TIP 115	PNP	60	60	5	2 A	50 W

Fig. 3-8. Parameters for add-on transistor for the voltage divider. Specs shown are at 25 degrees Celsius.

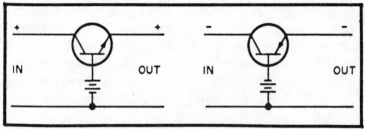

Fig. 3-9. Schematic for basic instant regulator circuit, showing both positive and negative voltage hookups.

source of DC voltage higher than the voltage you need, you can make the regulator by adding only two components, a transistor and a battery. The circuit is shown in Fig. 3-9 for both positive and negative output supplies. As you see, the battery supplies base current to the pass transistor, which acts as a variable dropping resistor. The battery drain is equal to the current supplied divided by the transistor gain. The output voltage is equal to the battery voltage minus the base-emitter drop.

If closer control of the output voltage is needed, it can be obtained by adding series diodes as shown in Fig. 3-10. Also, if better regulation is needed, it can be obtained by using a pair of transistors in the Darlington connection, as shown in Fig. 3-11. This connection is also worthwhile if the supply is to be operated for any length of time, since the battery drain becomes very small.

Suitable sources for the unregulated voltage are a battery charger, a car battery, or an old filament or bell transformer

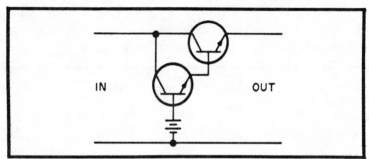

Fig. 3-10. Schematic for instant regulator for increasing or decreasing the voltage.

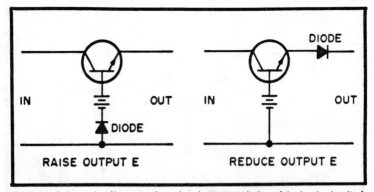

RAISE OUTPUT E   REDUCE OUTPUT E

Fig. 3-11. Schematic of improved version, better regulation of the basic circuit of Fig. 3-9.

with a series rectifier. In many cases, the charger or series rectifier does not even need a shunt capacitor to reduce ripple.

IC PROTECTION CIRCUIT

This IC-saving circuit is so simple it may appear too obvious to suggest; however, should a regulated power supply pass transistor short-circuit, $100 worth of ICs may easily be destroyed by excessive voltage, for example. The cost of new ICs, plus the often tedious task of troubleshooting and replacement, emphasizes the need for such positive protection.

Essentially, a selected zener diode is wired internally across the regulated power supply output. A typical applica-

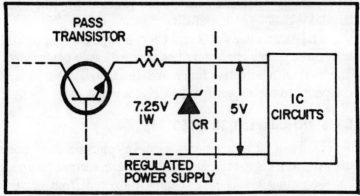

Fig. 3-12. Schematic of IC protection circuit. This little add-on protects your chips by not letting the supply exceed 7.25V.

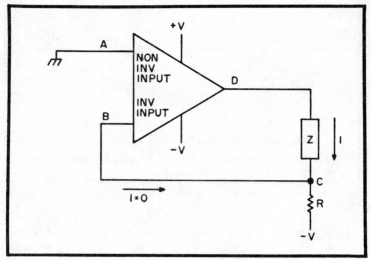

Fig. 3-13. Schematic of any operational amplifier used as a constant current source.

tion for protecting the popular 7400 TTL IC series is outlined in Fig. 3-12. This particular IC family has a power supply span of 0V to a 7V. The device design supply voltage is 5.5V; therefore, during normal power supply operation and the protective circuit draws little current.

With abnormal supply voltage, due to defective regulation or any other cause, the excessive voltage is dissipated in R and the zener diode. The load devices are protected because the applied voltage cannot exceed 7.25V. The power supply transformer and rectifier are protected by the combined dissipations of R and CR.

The exact values of R and CR are dependent upon load and power supply characteristics and are easily determined. The cost of a selected zener and resistor is insignificant, compared to the expensive devices they may save.

CONSTANT-CURRENT SOURCES

Constant-current sources are developing more and more uses today. Nickel-cadmium batteries are current charged. Light-emitting diodes are usually powered from current sources. Oscilloscope circuits use current sources charging a

236

capacitor for generating the sweep. State of the art ICs can provide precise control of current.

The operational amplifier is an excellent current source provided a stable supply voltage is available. To understand the current source it is only necessary to know that an ideal op amp has infinite gain, infinite input impedance and its differential input voltage is equal to zero. This is not exactly true but practically speaking we can accept it as such. Refer to Fig. 3-13. Since the voltage at point A is zero, the voltage at point B (and C) must be zero. No current flows from points B to C. In other words, the output of the op amp reacts such that it keeps point C at zero. A constant voltage appears across resistance R, and thus a constant current flow thru it (and Z) regardless of the value of impedance Z. For instance if the supply is ±15V and R = 15K then current I = 1mA. The current is constant within 1% or better as impedance Z changes from 0 to 12K or so. Naturally this circuit will work with any op amp.

A μA709 could be used for instance but then frequency compensation capacitors would be needed. The μA741 (Fig. 3-14) has the additional advantage of being short-circuit protected. To design for a particular current simply let $R = V/I$. Remember that voltage V is the greatest amount of voltage available to produce current I thru load Z. Note: The μA741 will handle up to approximately 25 mA.

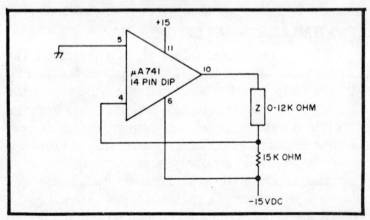

Fig. 3-14. Schematic of a μA741 op amp being used as a constant current source.

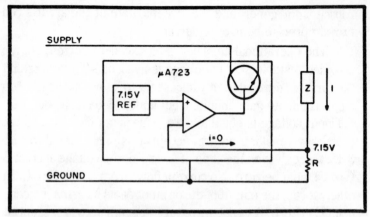

Fig. 3-15. Schematic of a more advanced form using the μA723 op amp as a constant current source. The voltage at R is held to 7.15V.

A more advanced current source employs a μA723 Fairchild voltage regulator. This circuit does not require a stable voltage supply. The IC has a built-in 7.15V reference. Refer to Fig. 3-15. This simplified diagram shows how the IC acts to hold the voltage across resistance R at 7.15V and thus cause a constant current flow thru load Z. In Fig. 3-16, resistance R = 15K and current I approximately equals 0.5 mA for Z from 0 to 50K or more. Be careful not to use a supply voltage larger than 40V DC. This is maximum for the μA723. The performance of this circuit is excellent with current constant well within 1%.

TWO-TERMINAL CURRENT LIMITER

Many power supplies will not survive a short circuit. The reason for this is that in the low-voltage supplies used for transistor circuit work, a series transistor is used to regulate the output voltage. If an unusually large current is demanded from the supply, and there is no current limiting, the peak current rating of the regulating transistor may be exceeded.

Other than the sensitivity of some power supplies to overload, experimental circuits are also subject to damage if for some reason excess current is allowed to flow.

A fuse alone will not necessarily provide the needed protection for current sensitive semiconductors.

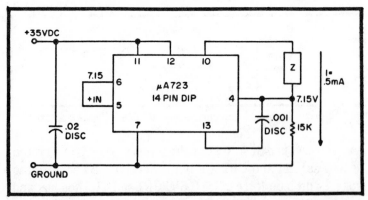

Fig. 3-16. Schematic of a μA723 chip hooked up as a constant current source with a 15K load, producing 0.5 mA of constant current.

The simple two-terminal current limiter shown in Fig. 3-17 will give instantaneous limiting for those slips of the probe, sudden shorts, etc. The use of a fuse in series with the limiter will reduce the need for a heat sink on the transistor Q1.

The limiter is placed in series with the line so the current is from collector to emitter. Though shown polarized in Fig. 3-18, the limiter may be used with either polarity supply.

In order to design for a particular maximum current, select R_e such that

$$R_e = \frac{0.6V}{I(amps)}$$

and make R_b about ten times as large as R_e. CR1 and CR2 are silicon diodes such as 1N4002, 1N645, 1N2070, or just about any diode capable of at least 100 mA. Q1 is a silicon transistor capable of the current to which the limiter is designed.

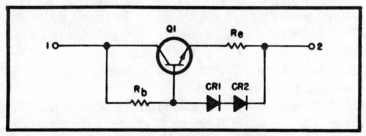

Fig. 3-17. Schematic of two-terminal series current limiter.

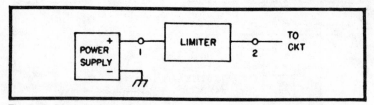

Fig. 3-18. Block diagram of two-terminal current limiter in typical application.

It works like this. R_b is small enough to keep Q1 saturated as long as CR1 and CR2 are not conducting. As soon as the voltages drop on R_e reaches 0.6V, CR1 and CR2 conduct and regulate the output current by regulating the base voltage on Q1.

MICROFARAD MULTIPLIER

A power supply is easier to filter if the load current to be drawn is small. As the load current increases, the size of the filter capacitors must increase if the ripple is to remain the same. It would be nice indeed if a circuit could be devised that would be able to magnify the effect of a small capacitor and make it act the same as a larger one. For one thing, capacitors of the electrolytic type are expensive, and for another, they are rather large.

The circuit described here does just what we have been discussing, and has been called a capacitor multiplier. It works so well that usually even the filter choke may be thrown out, and in high-current drain supplies too.

In the circuit diagram of Fig. 3-19, notice that the transistor is connected from the output of a conventional silicon rectifier bridge to the load, represented here by a resistor. The transistor base is biased by a 510Ω resistor and a 1.6K resistor, connected in a divider across the supply potential. The key here is 1000 μF electrolytic C2 from base to the plus side of the supply. This is the capacitor whose value the transistor multiplies. The multiplication factor is the DC gain (beta) of the transistor.

A simple formula can be used to describe the action of the circuit. $C_{eff} = C_{actual} \times B$ where C_{eff} is the effective filter capacitance across the output of the supply, C_{actual} is the value of capacitor in the transistor base circuit, and B (beta) as we have mentioned is the DC gain of the transistor.

The DC gain of a typical power transistor at say 0.5A is 50. This means that a 1000 μF capacitor in the base circuit will provide the same action as a 50,000 μF filter across the output of the power supply. If you don't think that's much, go to your catalog and price a 50,000 μF, 50 V electrolytic. If you buy one, try to find room for it on that crowded chassis.

CHEAP ADD-ON REGULATION

Although a number of regulated power supply circuits have recently been published, all have required specialized components, some of which may be available only through mail order. Here is a circuit which any electronics enthusiast, who has accumulated the usual collection of unidentified transistors, diodes, resistors, and capacitors can build a good quality regulated power supply cheaply and in a short amount of time. This regulator can provide either fixed or variable voltage as desired from either a transformer-rectifier DC source or an ordinary car battery charger.

This regulator consists of an NPN transistor in common base configuration and a PNP power transistor in common emitter configuration. The current gain of the transistor pair is the product of the gain of each of the transistors, but the

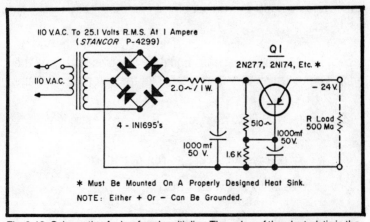

Fig. 3-19. Schematic of microfarad multiplier. The value of the electrolytic in the base circuit if the transistor is multiplied by the gain of the transistor.

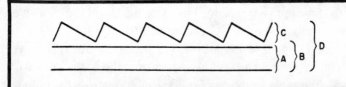

Fig. 3-20. Voltage waveform. (A) Reference voltage and regulator output voltage. (B) Minimum ripple voltage. (C) Ripple voltage. (D) Regulator input voltage.

feedback is such that the voltage gain is unity. As a result, the output voltage of the regulator is almost exactly equal to the reference voltage. If the reference voltage is pure DC, the output voltage will also be DC and independent of the regulator supply voltage so long as the reference voltage is less than the minimal ripple voltage (Fig. 3-20).

The basic circuit of the regulator is as shown in Fig. 3-21.

Transistor T1 can be practically any NPN transistor. This type can be identified with an ohmmeter because the base-collector and base-emitter resistances of an NPN transistor will be low when the positive lead of the meter is connected to the base. Conversely, the resistances will be high when the negative lead is connected to the base.

Transistor T2 should be a PNP power transistor mounted on an adequate heat sink. Most can handle a collector current of 2A or 3A. It can be identified, as it will have a low resistance between the base and either the collector or emitter with the negative terminal of the ohmmeter connected to the base.

The only moderately critical part in this circuit is capacitor C1. This capacitor should be large enough that the ripple voltage minimum is greater than the reference voltage by 2V or 3V. See Fig. 3-20B.

Max ripple = 1.5(RMS input) − 2.5 − reg output volts

or

= peak input − 2.5 − reg output volts

C1 must have a value greater than

$$C = \frac{\text{load current}}{120 \text{ (ripple volts)}}$$

Capacitor C2 is optional but it can improve the transient current regulation if the load current varies widely, as occurs in class B power amplifiers. It can have a value anywhere in the range of 50 to 500 μF.

R1 should have a value between 10K and 20K and it simply limits the emitter-base current of T1 should the reference voltage be present and the regulator supply voltage be off. R2 should be used either if C2 is used or if there will be times when the regulator will not be loaded. Select a value so that the current is between one-tenth and one-twentieth the maximum current output of the regulator.

There are a number of methods to provide a reference voltage, and all are adequate. Remember that the quality of the reference is directly reflected in the output of the regulator.

Figure 3-22 shows three methods to supply a reference voltage. Figure 3-22A shows a zener diode reference supply. R3 should limit the zener current to about 5 mA. If C3 is larger than 500 μF the reference voltage ripple will be less than 10 mV. (This can be calculated with the formula used above.) Figure 3-22B is a poor man's voltage reference. Each forward biased silicon diode will drop about 0.7V. So use as many as you need. Figure 3-22C indicates a simple method to provide a variable output voltage from the regulator.

After the regulator is constructed, check the output voltage with the reference. They should be the same. If the output voltage is higher, open the base circuit of T2. The

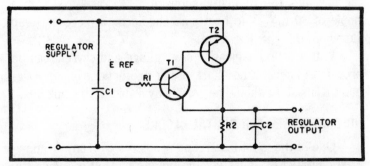

Fig. 3-21. Schematic of basic regulator circuit. The voltage gain of the transistor pair is near unity due to feedback. What goes in at the reference comes out.

243

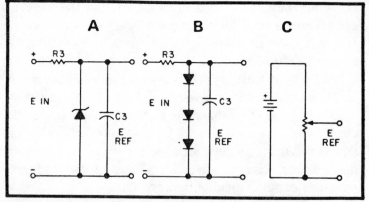

Fig. 3-22. Methods of supplying a reference voltage to the circuit in Fig. 3-21. (A) Fixed voltage by way of a zener diode. (B) Fixed voltage via the forward voltage drop of a diode string. (C) Variable voltage.

output voltage should then drop below the reference. If it does not, try another PNP transistor. If, on the other hand, the output voltage is low, there is a possibility that T1 should be replaced.

The voltage supply to this regulator is unfiltered DC and can most easily be obtained from an automotive battery charger, but a transformer with a full wave bridge or a full wave rectifier with sufficient voltage and current capability is fine.

It is suggested that you use a battery charger for a rectified AC supply which has an RMS output of 13V. The values of the components used were: C1, 2000 μF; C2, not used; C3, 500 μF; R1, 10K, R2, not used; and R3, 570Ω. With these values the regulator supplied 12.2V at 1.2A, with a ripple of 30 mV. The heat sink should be a piece of scrap aluminum 0.25 × 3 × 8 cm.

With a little imagination any voltage and current can be provided. There is no short circuit protection, but replacement transistors should be readily available in the junk box.

HOLDING POWER FOR DIGITAL CLOCKS

One of the most frustrating problems with any AC powered digital clock is its volatile memory. That is, when the power fails for even an instant, the memory of the displayed

244

time is destroyed, and when power is restored, the clock will not display the correct time. With some MOS chips, the clock will reset to 12:00:00 or 00:00:00, while with others any time may be displayed, even impossible times such as 28:42:00. Several methods of preventing this situation are available: (1) use a crystal oscillator timebase and battery power; (2) use the AC line for timebase and primary power with battery power to prevent memory dump when the AC power fails; and (3) use the AC line with capacitor backup as a power source when the AC power fails for momentary retention of clock memory.

Obviously, the first method is the most desirable setup since the clock can then be used in a car, boat, or any portable requirement. It is the least desirable from the standpoint of cost and complexity. A partial diagram of such an arrangement using the 53XX series of clock chips is shown in Fig. 3-23.

The second method is a relatively uncomplicated arrangement which prevents loss of memory and, depending upon the length of time that the AC power is off, will provide good service. The clock is triggered during the time the power is off and will show a loss of time when the power returns. Also, during the power off time, the display will not be illuminated. Since the displays draw the majority of power, only the clock chip is battery powered during the AC power failure. Figure 3-24 shows a partial diagram of the battery powered arrangement. Either dry batteries or rechargeable batteries can be used and are depicted in Figs. 3-24A and 3-24B respec-

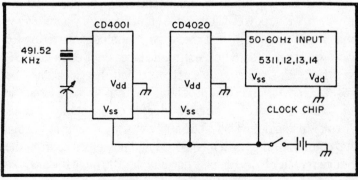

Fig. 3-23. Typical clock circuit using the 5300 series of chip.

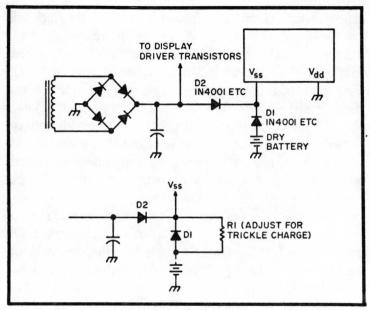

Fig. 3-24. Partial circuit diagram of typical battery-powered digital clocks. (A) Dry battery memory backup. (B) Trickle charger battery backup.

tively. In effect, the battery "floats" across the rectified AC power source and is switched on by diode D1 upon failure of the AC power. Note also the use of diode D2, which isolates the display power from the battery. If this diode is not installed, the battery will be quickly discharged by the display current drain. In Fig. 3-24B, a method is shown which will keep the rechargeable nicad battery on trickle-charge from the AC power source.

The third solution is one of the best solutions to the problem. Most power failures are of a momentary nature (less than 2 or 3 seconds), and these are the most frustrating for clock owners. In some cases, the failure is not even noticed, yet the clock's memory is destroyed and the clock must be reset. To overcome this problem, you can use the circuit in Fig. 3-25. Notice that the only addition to the standard clock circuit is one diode. This diode acts as a switch. During normal AC operation, the switch is closed and power flows from the bridge rectifier to the display directly, and through the diode to the filter capacitor and the clock chip. Upon power failure, the diode switch is reverse biased by voltage stored in the filter capacitor and the display is not illuminated. However, power from the capacitor flows continuously into the clock chip for several seconds and the memory is retained. Since the MOS chips draw only about 8 mA and will operate down to about 7V to 8V, the capacitor can supply power for several seconds. By using a 14V supply and a 3000 μF capacitor approximately 5 seconds will pass before memory destruction occurs. This is by far the least complex and least expensive method of improving the reliability of digital clocks.

One other tidbit of information concerns the 50252 clock chip sold by Radio Shack and other dealers. The chip requires that you depress both the tens-of-minutes and hours time set switches to advance the minutes display. This is somewhat cumbersome and more times than not you wind up advancing the tens-of-minutes digit or the hours digit because the switches were not depressed simultaneously. The solution involves a third switch and two diodes (1N4148s or similar) and is shown in Fig. 3-26. The added switch now advances only the minutes display with no worry of accidentally advancing any of the other displays. This particular clock chip has several desirable features such as 24 hour alarm, 10 minute snooze alarm and a built-in tone generator for the alarm. However, it does have one disadvantage which is not immediately obvious—it cannot be operated in the 24 hour mode on 60 Hz power. This feature is a matter of individual preference and was somewhat of a disappointment. Why the chip functions like this is not known since the alarm feature always

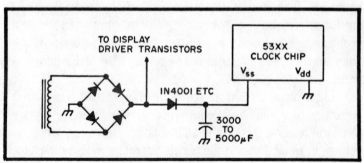

Fig. 3-25. Capacitor backup for short-term memory protection of digital clocks.

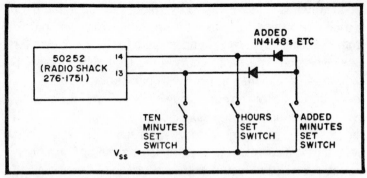

Fig. 3-26. Special setting circuit for 50252 clock chip so that the minutes only advance.

operates in the 24 hour mode regardless of the input power line frequency (50 to 60 Hz); however, the display will only operate in the 24 hour mode when the input is 50 Hz.

CAPACITOR REJUVENATOR-LEAKAGE TESTER

The explosion of an electrolytic capacitor is not only unnerving, it is downright dangerous. True, there are not many capacitor explosions, but filter capacitors with deformed electrolytes are a major source of smoke during the initial testing of power supplies. The instrument described here can pay for itself by preventing the destruction of only a handful of electrolytics.

After a nerve shattering explosion of electrolytic salvaged from some unremembered defunct gear, it was decided to prevent recurrence of this episode.

First, the cause of the smoke and explosions needed to be known. That meant searching through theory books looking for information on a subject seldom mentioned. Such a problem does not occur in industry or repair shops since new units are always used. Here is a summary of available information:

Electrolytic capacitors are made of two plates of metal foil separated by a thin film of chemicals called electrolyte. Under electrical tension this film is an insulator. However, when no voltage is applied to the foil plates for many months, as happens when a unit is stored in a junk box or is in unused equipment, the film tends to "deform." This means it is no longer a good insulator. When voltage is applied, current is permitted to pass through the film. Heat is generated in the resistance. The heat becomes great enough to vaporize some of the electrolyte if the current is high. Pressure builds up, and something's got to give. A unit which has one end sealed with a rubber-like compound will seldom explode, but will break open and fill the room with stinking yellow smoke. Capacitors firmly sealed will give more thrilling results.

The electrolyte can be reformed by applying a gradually increasing voltage and keeping the current below about 1 mA. This voltage can come from a capacitor tester, since most models have variable voltage provisions. The drawback to this approach is that the testers usually have a spring releasing safety switch that applies the voltage. To rejuvenate a capacitor then means sitting with thumb on switch or laying the tester on its back and weighting the switch with a spare transformer.

Neither of these alternatives is particularly attractive. It was determined that it would be fairly simple to design a gadget that could provide a variable voltage and a current limited to a safe value for reforming the electrolyte.

To start with, we need a voltage source that will provide adequate potential to reform most capacitors in common use. A voltage doubler working directly off the power line would be cheap and simple, but also dangerous, since one side of the line is grounded. This is easily remedied by using an isolation transformer. A full-wave voltage doubler gives slightly more than twice the transformer RMS output voltage when the drain is light (see Fig. 3-27).

A voltage divider of five 22K resistors will draw about 3 mA. At the save time, multiples of 60V are available so they can be applied to the capacitor.

An 82K resistor is used to limit the current through the reforming capacitor. Some means is needed to know when the capacitor can withstand the applied voltage. A meter would be nice, but expensive. A neon bulb connected across the current-limiting resistor will be extinquished when the current is less than about 1 mA.

Since the voltage was available, and the drain was light, it was decided to include a leakage checker for paper capacitors. Only four additional parts were required: a 200K resistor, a NE-2 bulb, an SPDT switch, and J3. This makes possible a quick and simple, but positive leakage test for tubular capacitors.

Construction of the rejuvenator-leakage tester is straightforward. The voltage doubler was assembled on a terminal strip before mounting it in the box.

The neon bulbs used as indicators were GE plastic encased units that mount snap-in fashion without need for other hardware.

Test leads using insulated clips are a must since there is as much as 300V applied to them when the instrument is in operation.

An accessory socket was added so that power from the 6.3V winding and voltage doubler could be used to operate one-tube projects.

To rejuvenate a capacitor, connect its negative lead to the ground jack and the positive to the voltage divider terminal. The voltage level should be at zero during connection. After it is properly connected, advance the voltage level control to the first position. The indicator light in the circuit should light briefly. Leave the switch in this position for a few minutes, then advance it to the next level. Repeat this procedure until the capacitor is charged to its rated voltage. With the five steps used in the original model each advance represents an increase of 60V. Before attempting to disconnect the rejuvenated capacitor, return the voltage level control to the zero position. This shorts the capacitor, thus discharging it.

The instrument produces only a little over 300V. However, no capacitor reformed to this voltage has broken down in any circuit using up to 400V.

To use the leakage checker section, simply connect the capacitor to the appropriate leads and depress the SPDT switch, which is a spring return switch. The leakage test lamp should blink only once. If it blinks on continuously, the unit under test is leaky. If it fails to blink at all, the capacitor is

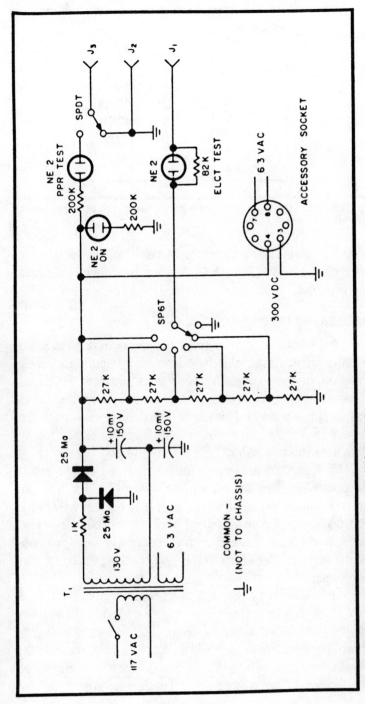

Fig. 3-27. Schematic of capacitor rejuvenator-leakage tester.

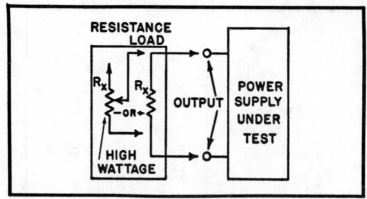

Fig. 3-28. Block diagram of a simple method of testing a power supply with a high-wattage resistor.

probably open. The range of values over which the open test is effective is 0.001 to 0.5 μF. Releasing the switch discharges the capacitor.

VARIABLE DC LOAD

Quite often electronic construction projects go to great lengths to describe the functions of components, their minimum or maximum allowable values and how the circuit works, but end up skimping on explanations about the power supply section. More likely than not they merely state the required voltages and currents and leave the power supply design up to the builder.

Most electronic manuals furnish data and design details for power supplies. If you are the type of builder who can afford to go and buy all new components to build exactly to the handbook design then you won't have any trouble. However, in this world of surplus and used TV set power supply components, most builders turn to the junk box first for the power supply parts.

Since it is rare to find the exact components, even in the best stock junk box, the builder usually ends up with a power supply constructed with components quite a bit different than those specified. We know it will work but *how well* will it work? Naturally, attaching it to the piece of gear it will power will be the "acid test," but why compound problems? Why try to test and debug a new piece of gear with a power supply that may be bugged because of inadequate regulation or filtering. A power supply defect can really hide itself and show up as a bug on the gear being tested—like distortion in a linear amplifier because of poor power supply regulation.

Why not test the power supply off line before using it? Here's how.

If you have ten or twenty assorted high wattage resistors or a 200W variable resistor in your junk box and don't mind the inconvenience of dropping power to change the resistor or vary the tap then do so as indicated in Fig. 3-28.

If you have a couple of old transmitting tubes with high plate dissipation, a filament transformer to match, a couple of low-wattage variable resistors (one being a potentiometer), a DC milliameter, and a small power transformer for bias, plus want complete flexibility and ease of control, then do as indicated in Fig. 3-29.

Since a couple of old VT-4-Cs (211) with big graphite plates rated at 100W were in the junk box, the procedure in Fig. 3-29 was chosen. This decision was also prompted after burning several fingers a couple of times changing hot resistors as in procedure shown in Fig. 3-28. Actually, any tubes will work. The decision as to type should be made based on plate dissipation rating. Pick the tubes with the highest rat-

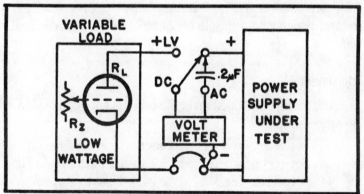

Fig. 3-29. A slightly more complex method of testing power supplies with a handful of surplus components.

ings. The maximum loading and power consumption by the unit will be determined by the plate dissipation of the tubes used.

How It Works

Figure 3-30 shows the diagram of the variable load. The load consists of tube power dissipators and a bias power supply with a means (R2) of controlling the bias and therefore current drawn by the tubes. A DC milliammeter (M1) is inserted in the cathodes of the tubes to measure the current through the load. The meter has a basic $0 - 100$ DC mA range that is extended to $0 - 500$ mA by switching (S2) a shunt (R3) across it to test heavy duty power supplies. A voltmeter is connected between the input terminals $-V$ and $+LV$ or $+HV$ to simultaneously measure power supply output voltage or ripple and current. The 100Ω 10W resistors in the plate circuits of the tubes are necessary to assure stable tube operation. The resistor in the plate lead from the $+HV$ is necessary if higher than rated voltages will be tested.

As stated earlier the current drawn by the tubes is controlled by R2, which varies the bias on the tubes. Basically, a tube is cutoff (drawing no current), when the grid is more negative than the cathode. A tube is conducting (drawing current), when the grid is more positive or moved closer to the cathode potential. The degree of conduction or current draw depends on the bias setting between these two extremes, hence, the principle of operation of the variable load unit.

Adjustment

The cutoff bias condition of the tubes is set by the following procedure:

1. Set slider on R1 to midrange.
2. Set potentiometer arm R2 to full minus end. The potentiometer should be wired so this condition exists when the arm is rotated fully counterclockwise.

3. Connect the unit to an external power supply. This supply should be one with a highest output range available which can be used with the type of tubes used in the variable load.
4. Turn on the variable load.
5. Turn on the test power supply.
6. If milliammeter M1 shows any indication, turn all power off and move the slider on R1 towards the plus end of the resistor. If the meter does not move go to step 7. Turn on power, repeat steps 4 and 5, and again check the milliammeter for a reading. Repeat this step until R1 slider is set at a point that current just stops flowing.
7. If the milliammeter M1 does not show any indication turn all power off and move slider of R1 slightly towards the minus end of the resistor. Turn power on, repeat steps 4 and 5, and again check the milliammeter reading. Repeat this step until R1 slider is set at the point where current starts flowing, then

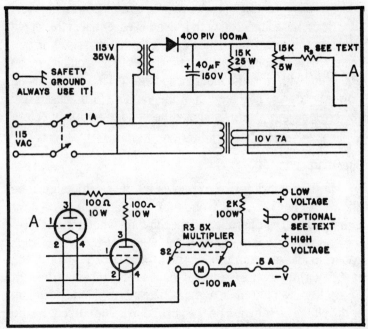

Fig. 3-30. Schematic of the variable DC load circuit—the most practical method.

back off R1 slider to the point where current just stops flowing.

8. Steps 6 and 7 will set the bias point of the variable load so that its maximum range of loading can be utilized.

9. Insert a milliammeter between the arm of R2 and the grids of the tubes and measure the current drawn when R2 is set for full conduction. If the value is higher than that allowable for the 5W rating of R2, insert a resistor at RS to bring the current back down. The slider of R1 should not be set any closer than 15% of its total value to the minus end of the resistor. This will limit the tube grid current when R2 is set for maximum conduction at the plus end and help prevent R2 burn out.

Grounding

Notice in Fig. 3-30 that the only thing grounded to the chassis is the ground conductor of the three-wire AC line cord. This will allow testing of minus as well as plus power supplies. The three-wire line cord with ground is essential in this application in the event of a short circuit to the chassis. For that matter it is recommended that all electrical equipment be equipped with three-conductor power cords. If grounded power cords are not used then the variable load chassis and the power supply chassis should be bonded together before running any tests.

Operation

In the following two examples the test setup should be shown in Fig. 3-29. The voltmeter should be connected for either AC or DC operation, depending on whether measurement of regulation or ripple is the object of the test. Graph paper should be used to plot the progress or results of the test. It's a lot easier to see what a series of numbers represent when they are plotted on a graph. The graph should be set up with voltage increments on the vertical axis and current on the horizontal axis. The steps or size of the increments will generally depend on the two extremes of the measurement. Try to keep the graph spread out for accuracy.

Once the equipment is set up power is turned on and load control R2 is turned clockwise until the amount of current representing the first current increment is indicated on the milliammeter. The voltmeter is then read and the indicated voltage is logged as the first spot on the graph. R2 is then advanced to the next current increment and the voltage is again read and logged. This process is continued through the last current increment at which time a line should be drawn connecting each spot on the graph paper, completing that part of the test. At this point a component can be changed and the whole sequence runs again to see if an improvement was realized.

These two examples will give you an idea of how it works:

1. **Regulation.** The graphs in Fig. 3-31 were made by testing a power supply using a transformer with an unknown current rating. Two unmarked chokes, L1 and L2, were tied in the power supply to see which offered the best regulation. A DC voltmeter was

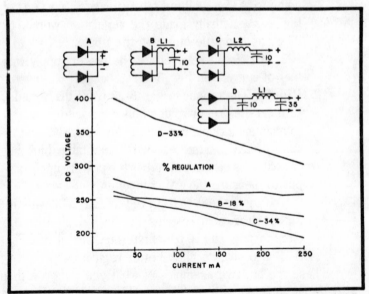

Fig. 3-31. Test results, obtained quickly and easily, illustrate power supply performance under load.

257

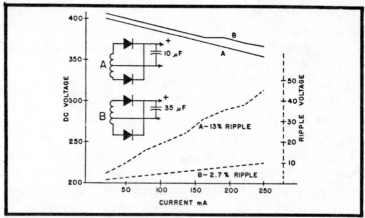

Fig. 3-32. As power supply load increases, voltage drops together with an increase in ripple.

connected across the power supply output to measure the voltage under the different current settings.

Examination of curve B and C of Fig. 3-31 shows that the choke used in curve B has 16% better regulation (AC ripple was the same in both cases). The addition of a 10 μF input filter capacitor and going to a 35 μF output filter capacitor, curve D of Fig. 3-31, actually made the regulation worse, to 33% and had little effect on the ripple.

The circuit selected for the power supply was that of curve B in Fig. 3-31.

2. **Ripple.** The graphs in Fig. 3-32 were made to determine the proper value of filter capacitor required to maintain a ripple constant of less than 3%. In this case an AC voltmeter, with a suitable blocking capacitor, is connected across the power supply output to measure the AC component.

It can be seen that curve B of Fig. 3-32 with the 35 μF filter capacity gives 11% less ripple than the circuit using only 10 μF of filtering. Circuit B of Fig. 3-32 also gives slightly better regulation.

These are but two examples of what you can do with a variable DC load. As with anything else, its total effectiveness will be up to the ingenuity and inventiveness of the user.

Chapter 4
High-Voltage Power Supplies

Here in this chapter we have used to term high voltage to mean anything over 50V. Most of the power supplies in this chapter will apply to those circuits powering vacuum tubes. Very few solid-state circuits these days require more than 50V, but there are those few exceptions that you may run into from time to time. To really start off with high voltage, the first power supply in this chapter generates 3000V. From that point on things taper off a bit. Several of the high-voltage supplies here are intended for use with an SSB transceiver, but don't let that restrict you—use them as you will.

3000V DC SUPPLY

Have you ever built a piece of gear with available parts and done an immaculate job, only to have a transformer or component go bad after a period of time? When you go to purchase a replacement you usually find it is sold out or no longer available. So you shop and shop, trying to find something that will do the job and also fit into your cramped dimensions. Perhaps you'll be lucky, or you'll end up rebuilding.

The described 3000V power supply (Fig. 4-1) incorporates rugged design specifications coupled with generous di-

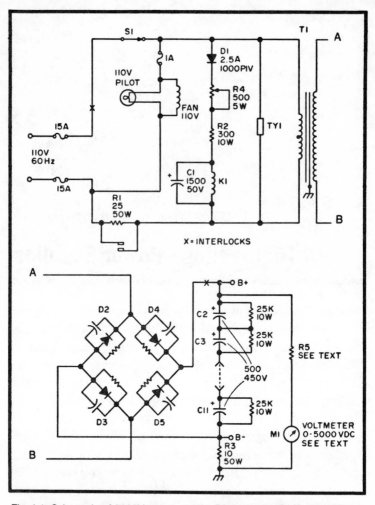

Fig. 4-1. Schematic of 3000V power supply. Diode stacks D2 through D5 are constructed of eight 2.5A 1000 PIV series connected rectifiers each. Shunted across each diode is a 470K 1W resistor and a 0.01 μF 1000V disc capacitor. Capacitors C2 through C11 should be 500 μF with a minimum voltage rating of 450V DC. K1 is a P & B type PR3DY 24V DC coil with 25A contacts. Transformer T1 has a 2200V RMS secondary with a 500 mA minimum rating. The thyrector is a GE 6R520SP4B4.

mensions that provides versatility in accommodating transformers and related components found on the surplus market.

Construction

The high voltage power supply is easily constructed, as all mechanical work can be performed with a metal munching tool, pop rivet gun, good soldering gun, and ordinary hand tools. An electric drill with variable speed control will save much time.

The main aluminum chassis and front are 33.02 cm × 43.18 cm × 7.62 cm (standard) and the back aluminum wall has a 2 cm inside lip at the top and bottom for attachment to the main chassis and cover. The front chassis and rear wall are attached to the main chassis by generous use of pop rivets. The main chassis is reinforced on the bottom with a thick steel plate with one caster at each corner and one in the middle to support the weight. The line cord is fed to the rear through a steel conduit.

The cover is manufactured by hand bending a sheet of aluminum to tightly fit the chassis assembly and is held in place by sheet metal screws. Right angle aluminum brackets were installed on the back plate along the sides to accommodate the fastening of the cover. Air is exhausted by mounting home air vent assemblies on the sides of the cover. A local lumber yard has the vents.

Before attaching aluminum to aluminum, rough each contact surface with fine sand paper to ensure a good electrical connection. Also, connect each chassis and the back together electrically with copper braid.

The front of the supply contains a voltmeter, on-off switch, and pilot light. The rear of the supply is designed with safety in mind. The B+ B − connections are in a minibox with two grometted holes in the bottom. The large insulated feedthrough was fitted on a small Plexiglass sheet and the hole in the aluminum made extra large to prevent high voltage breakdown. High voltage cables should have a minimum rating of two to three times the DC output voltage.

The diode stacks are made by mounting sight diodes on four prepunched epoxy boards. The insulated spacers for the boards are nothing more than self-tapping plastic expansion tubes, available at most hardware stores. The boards are connected to the spacers and the spacers to the chassis by nylon screws.

The filter capacitors are mounted in holes drilled in Plexiglass with a hole saw. The Plexiglass is held in place by self-tapping plastic expansion tubes and nylon screws. To prevent the capacitors from arcing to the chassis, the area below the capacitors has a sheet of punched epoxy paper board cemented to it; the cutout plastic circles are cemented to the bottom of each capacitor.

To keep air circulating, a small fan is mounted in the rear of the supply. The fan is fused, and the fuse is located in the front under chassis where it can be changed without removing the cover. The AC line is terminated in the front chassis at the switch and at this point the Thyrector is also located across the line.

Located at the lower right of the rear chassis is a heavy duty ground connector *which should always be utilized for maximum safety.*

All lettering is accomplished by the application of white dry transfers over black wrinkle paint.

Circuit

The circuit utilizes a full wave bridge rectifier circuit with a capacitor filter of 50 μF. This provides approximately 5% regulation with a 3K load. Ten 500 μF, 450V capacitors provide a total voltage rating of 4500V.

The high voltage diodes, capacitors, and transformer are protected from excessive current when the power supply is first turned on by series limiting resistor R1. The time delay for relay pickup is determined by R4, which adjusts the time required for C1 to charge and energize relay K1, which closes its contacts and shorts across R1. Too much delay causes R1 to overheat. One second proves satisfactory.

The supply also incorporates a voltmeter that measures the output voltage. An inexpensive meter can be utilized as the supply incorporates a resistor multiplier string to increase the range of the basic meter movement, *but never use a meter with a metal zero adjusting screw in high voltage circuits.* To choose the correct value of R5 for your meter, use the following formula:

$$R5 = \frac{\text{full scale desired}}{\text{meter reading in amps}}$$

A resistor is not a high voltage device; therefore, to achieve the desired resistance of R5 many series resistors must be used to handle the voltage. Ten 1M, 1W resistors in series mounted on a strip of epoxy board can be used, thereby distributing the voltage equally across ten resistors. The epoxy board is mounted to the front chassis on two ceramic insulators.

To protect the supply diodes from transients a Thyrector-diode assembly (TY-1) is installed at the line input. Also, each side of the line is fused to provide adequate protection to the supply and station line circuits.

Interlocks

All high voltage power supplies should contain an interlock or interlocks. Basically there are two types: the primary interlock and the secondary interlock.

The primary interlock is similar to the power cord assembly on a television receiver. When you remove the back of the set you open the AC line; the television cannot be energized by unauthorized personnel without a special line cord. See Fig. 4-2.

The secondary interlock normally, Fig. 4-3, shorts the secondary out thereby discharging any residual charge on the high voltage capacitor string, thereby also protecting the user from electrical shock due to an open bleeder or equalization resistor.

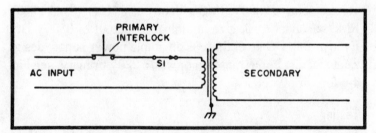

Fig. 4-2. Primary interlock. When removing the cover or panel of the supply, the primary interlock must have ample current carrying capability.

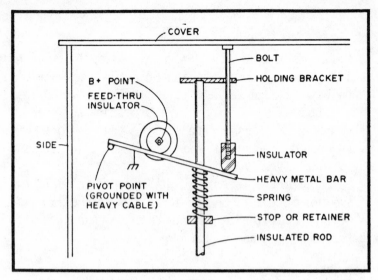

Fig. 4-3. Secondary interlock. Removing the cover permits the metal shorting bar to move up and contact the B+ point, thereby shorting any dangerous voltage to ground. As long as the cover is removed the B+ point will be grounded. This assembly must be mechanically strong and not subject to movement or bending.

Neither a primary nor secondary interlock alone will give 100% protection, but utilization of both in one supply will come close.

In essence, in respect to safety, it can be said that a power supply that does not break down requires minimum service; therefore, the best protection is to build high voltage supplies with generously designed safety factors.

Transformers

This supply can accept transformers with a secondary voltage of up to 2500V RMS with no design changes. A 2500V RMS secondary will give an unloaded output of 3500V DC. Even at this DC level there is an ample power supply design safety factor. For a transformer with a 220V primary, see Fig. 4-4 for wiring details.

Testing

When the circuit is completed, check the wiring and look for any possible short circuits. Between ground and any positive voltage points, look for a minimum of 3 cm separation when the insulation is solely air.

Also, inspect each electrolytic capacitor to make sure none of the exhaust ports are obstructed by construction. A defective electrolytic or an electrolytic with a plugged or blocked exhaust port can explode violently.

Before energizing the circuit, review the basic rules of safety.

1. Never bypass an interlock.
2. Fuse the circuit properly.
3. Never operate or test the supply with the cover removed or high voltage terminal exposed.
4. Make sure others in the household are aware of the location and operation of the master power cutoff switch so they can disconnect circuit from line in an emergency.
5. Label *all* high voltage points and equipment as such: DANGER—HIGH VOLTAGE.

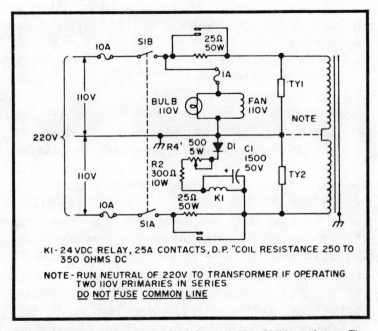

Fig. 4-4. Alternate 220V primary circuit for use with a 220V transformer. The components are similar to those used in the original circuit except two thyrectors are used and the relay K1 is a double-pole type.

6. Voltmeter should read zero and main AC line should be disconnected before removing cover or changing high voltage leads.

7. Make sure the family members know the basics of artificial respiration. Many shock victims die of suffocation before professional help arrives.

8. If you don't understand something, get the facts before proceeding.

9. Properly connect the power supply to a good and permanent earth ground.

Even when the circuit has been inspected and rules followed, there is the potential danger of defective new or used components. A 3000V DC supply with a 50 μF filter is a lethal device. Always assume that *all* points in a circuit of this type are dangerous and proceed with that in mind.

ADJUSTABLE 230V TO 350V SUPPLY

Everyone who has ever built a power supply has had the sad experience of coming out with a voltage which was too high or too low for the project at hand. The usual practice is to try changing from a capacitor input filter to a choke input, or the reverse, in order to raise or lower the available voltage. Sometimes this works if the voltage requirement is not too critical. Other times it doesn't, and we resort to expensive, power consuming, and poorly regulated voltage dividers.

As an example of a much more satisfactory method, here is a circuit in which a TV power transformer output is bridge-rectified for high voltage and the low voltage is obtained off the center tap of the same transformer. The high voltage came out exactly right; the low voltage, which was to have been 250V under a 100 mA load, came out at 230V with a choke input filter. An attempt at changing to a capacitor input failed miserably; 350V was much more than the circuitry could endure. It looked like time for the 20W resistors and all of that nonsense.

It would seem from the above that what was required was a filter which would operate somewhere between the choke and the capacitor input conditions, and this was what was finally used. In the figure you will note a resistor in series with

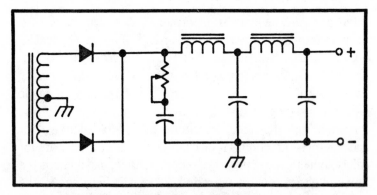

Fig. 4-5. Schematic of adjustable 230V to 350V supply.

the positive lead of the input filter section. This resistance, when fully in the circuit, seriously limits the ability of the capacitor to charge and as a result the output voltage is reduced almost to the value of a choke input. On the other hand, when the resistor is reduced in value, the input filter charges to a more normal voltage and the output reflects this change. As may be seen from these two examples, it is possible to adjust the output voltage by merely adjusting the series resistor for the desired voltage at the required load.

It would be desirable to substitute a rheostat or adjustable resistor for setup purposes and replace it with a fixed resistor when the exact value is known.

It is suggested that values below 1000Ω would be most useful. In the example cited above a 500Ω, 10W resistor was used. Try it, it works like a charm.

800V & 250V SUPPLY

The power supply described for the Heath HW-12, 22, and 32 SSB 200W PEP transceivers. The requirements, however, are representative of many on the market. It delivers about 800V DC at 250 mA peak, 250V DC at 100mA, −124V DC for grid bias, and 12.6V AC filament voltage. The bias features zener regulation, and the entire supply proves very satisfactory in service.

Power transformer T1 was acquired as new surplus. It is rated at 800V CT at 300 mA DC. The filament windings are rated at 6A. The bias transformer, T2, is simply a 6.3V filament transformer wired backwards. If this is not available, a 1:1 ratio 120V transformer could be used across the AC line. The high voltage portion is a full wave bridge rectifier, using twelve 600V PIV silicon rectifiers. It is filtered by a single L-filter consisting of L, C3, and C4. R2 and R3 equalize the voltage across the filter capacitors. R4 is a bleeder resistor and has three functions: it discharges the filter capacitors for safety, it helps to regulate the output voltage, and it keeps the output voltage, when unloaded, from climbing above the capacitor voltage ratings. Because of the quiescent current of the final amplifier, very little bleeder current is required.

M1 is a 1000V voltmeter. A milliammeter in series with the load would have been more typical, but it was desired to monitor voltage regulation rather than load current. It is optional, and either, or both, can be used.

The low voltage, from the transformer center tap, after being well filtered, supplies the receiver section and the low level transmitter stages. L2 and L3 are also surplus units and are actually one tapped inductor. R6 is a dropping resistor to obtain the correct output voltage.

The bias voltage is half-wave rectified by CR13. R7 is a surge resistor, and protects the rectifier. CR14 and CR15 are 62V 1W zener diodes, giving constant bias voltage regardless of line variations. Very little output current is drawn from the bias supply.

When wiring the filament windings in series, a voltmeter across the output will read either zero or 12.6V AC. If zero is read, reverse one of the windings, so that they add rather than cancel.

One side of the primary winding is connected through the transceiver function switch to the line so that the power supply can be switched on with the transceiver. SW1 is a high/low power switch. Since the power supply is solid state, there are no filaments to warm up, and the full output voltage is at the terminals immediately upon switching on. This is very hard on the transmitter tubes, since their filaments have not yet heated. With SW1 open, there will be no voltage at the low

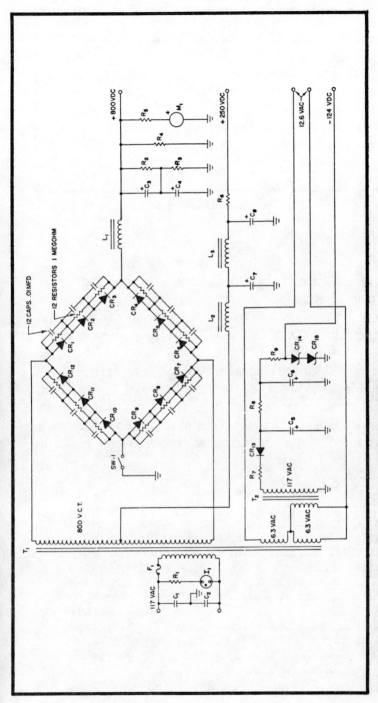

Fig. 4-6. Schematic of 800V and 250V supply that is ideal for an SSB transmitter.

269

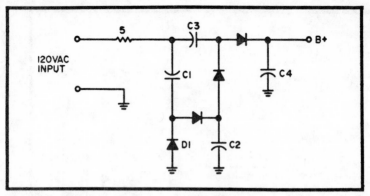

Fig. 4-7. Schematic of conventional half-wave quadrupler. The full output is across capacitor C4.

voltage output, and only half the normal voltage at the high voltage terminals. Full bias, however, will be supplied. After allowing filaments to heat for a minute or so, the switch can be closed, applying full voltage. This should lengthen tube life appreciably. While a SPST toggle switch works fine in the supply with no sign of arcing, it may be desired to use a ceramic wafer switch with a higher voltage rating. Of course, the same thing can be done with a time delay relay.

Much wiring can be eliminated, if desired, by purchasing a packaged bridge rectifier unit. These contain the silicon rectifier bridge, with all the required voltage equalizing resistors and capacitors, in a compact potted package with four terminals.

Parts List

C1, C2..................................0.001 μF, 1000V DC ceravic disc
C2, C4..............................30 μF, 500 V DC tubular electrolytic
C5, C8..............................20 μF, 450V DC tubular electrolytic
CR1, CR13 ....1N547, 600 PIV 750 mA silicon rectifier
CR14, CR15..............................1N3039, 62V 1W zener diode
F1..4A fuse
I1 ....NE51 neon indicator lamp
L1..8H, 250 mA filter choke
L2, L3..8.5H, 125 mA filter choke
M1..............................0 − 1 kV voltmeter, 1 mA movement
R1...270K, ½W resistor
R2, R3 ....150K, 2W resistor
R4...50K, 25W wirewound resistor

R5...1M, ± 1%, 1W resistor
R6.............................750Ω, 10W wirewound resistor
R7...15Ω, ½W resistor
R8...560Ω, ½W resistor
R9...4.7K, ½W resistor
SW1.............................SPST toggle or rotary switch
T1.......800V CT 300 mA, 6.3V AC 6A, 6.3V AC 6A transformer
T2.......................................6.3V AC 1A transformer

VOLTAGE SEXTUPLER FOR SSB

This circuit is an extension of a modified quadrupler. The resultant output voltage is more than six times the input voltage, about seven and a half, because the capacitors charge up to the peak of the AC voltage imposed on them, and after rectification more DC results than the average value of the AC input. Output DC is 900V with low voltage and bias.

Figures 4-7, 4-8, and 4-9 show how the circuit was developed. In fact, this voltage multiplication process could be extended an infinite number of times, but electrolytic capacitors have to be used in series, and when they do the effective capacitance is reduced. So there is a practical limit to this multiplication business, as the regulation begins to suffer as the effective capacity is decreased.

Voltage regulation is good, but not as good as with a quadrupler, due to the fact that the output capacitance is made up with C2, C4, and C7, three 300 μF capacitors in series,

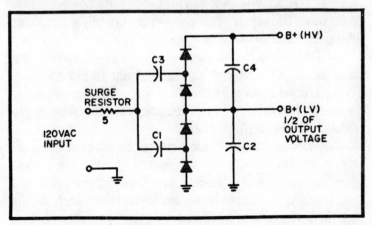

Fig. 4-8. Schematic of a modified quadrupler. By pitting capacitors C4 and C2 in series lower voltage rated capacitors can be used.

271

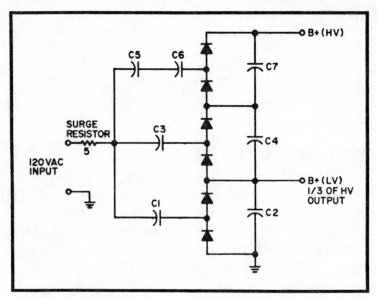

Fig. 4-9. Schematic of the basic sextupler. By extending the circuit in Fig. 4-8 with two more diodes and capacitor, we end up with a sextupler, multiplying the voltage six times. Two capacitors, C5 and C6, have to be used in series for the fifth multiplying step because approximately 750V appears at this point.

giving an effective 100 μF. C5 and C6 in series make up the capacitor for the fifth multiplication. Two 200 μF at 450V units giving an effective 100 μF. All capacitors are of the twist-lock, can type, except C5 and C6, which are tubular type. The 900V output drops about 80V under 250 mA drain, which is still under 10% regulation. Not bad for this type of power supply, or any other HV supply. The low-voltage tap only drops about 5V, under voice peaks.

A 5 × 9 × 2-inch aluminum chassis is used to mount the components. A front panel is used to mount a 4 inch speaker. The lone transformer is for the filaments, and is surrounded by the forest of capacitors. The schematic of the complete supply is shown in Fig. 4-10.

Mention should be made here of the surge resistor arrangement. It is an absolute necessity to protect the diodes, when the power supply is first turned on and the capacitors look like a dead short; however, after thirty seconds the resistor is shorted out by time delay relay K1, as the resistor serves no other purpose, and if left in the circuit causes a

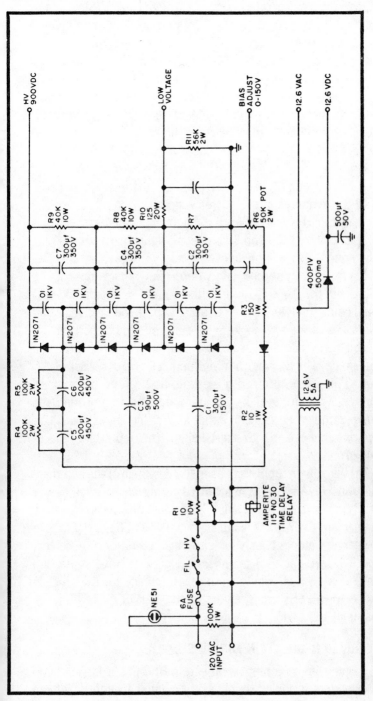

Fig. 4-10. Complete schematic of the sextupler power supply.

voltage drop as current is drawn through it. A 1V drop in the input causes approximately a 7.5V drop in the output voltage at the high-voltage tap.

Each diode has an 0.01 μF disc capacitor across it for transient protection.

All of the precautions should be used on this power supply since it operates with one side of the AC line grounded. However, nothing is to be feared if the AC line plug is inserted correctly. This can be determined by two different methods. The first is to run a ground lead from the power supply chassis to a good ground: cold water pipe, or a driven ground rod. With the switches in the power supply turned off, insert the AC plug into the wall outlet. If the neon bulb ignites, the plug is in correctly. If the neon bulb doesn't light, just turn the plug over, and you are in business.

The other method is to determine which one of the sockets in the AC wall outlet is the neutral or grounded leg of the incoming power. City electrical wiring codes specify that the larger of the two sockets be the grounded or neutral leg, but be sure and play it safe and find out for sure. After you determine which is the hot and neutral, use a polarized AC plug. This is the preferred method, and can be determined with an AC voltmeter by hooking one lead to a ground and plugging the other lead into either socket that gives you 120V. The one that doesn't give a reading is the neutral leg, therefore ground.

This power supply has been used with all of the commercial transceivers on the market requiring an external power supply, with excellent results. The value of R10 may have to be changed to drop the voltage on the low-voltage tap to give the correct amount for your particular brand of transceiver. The light weight, 5.5 pounds, and good voltage regulation make this type of supply an excellent choice for the modern day, compact SSB transceivers on the market today. And the cost isn't too bad.

IMPROVED REGULATION FOR HV SUPPLIES

Some improvement in the regulation of a high-voltage semiconductor power supply system with capacitor input can be obtained by shorting out the protective surge resistors placed in the secondary leads of the power transformer by means of a delay circuit. The amount of improvement, naturally, depends on the values of the resistors.

One method of accomplishing this objective is shown in Fig. 4-11.

When the plate switch in the primary of the high-voltage transformer is closed, the RC time constant provides a delay of about 1 second before the contacts of relay K2 close and short the surge resistors. Variation in delay from milliseconds to several seconds can be achieved by choosing values of R between $10K$ and 20K or substituting different values of C.

Since the time constant in the filter circuit of the power supply is so small, a delay of 1 second is sufficient to provide

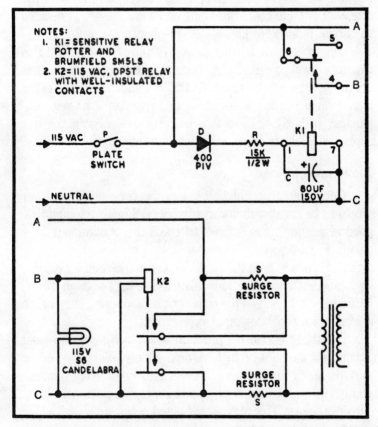

Fig. 4-11. Schematic of improved regulation circuit for high-voltage supplies.

275

ample protection in shorting out the surge resistors when the power is turned on. A 115V isolating transformer must be used between the plate switch and the diode if the 115V line polarity is not observed.

850V & 275V SUPPLY

A few years ago a new transmitter of commercial manufacture was acquired. This was a high quality unit, but came without a power supply, as do many medium-power transmitters. As a consequence, the first order of business was to build up a source of the various voltages required for the rig.

The first thought was to build a power supply like most of those being used today; that is, pack all the power capability possible into a very small size, using every trick to reduce the size, weight, and cost. But being conservative it was decided to pursue a different line of design philosophy.

The schematic of the supply is shown in Fig. 4-12 to illustrate two points. First, note that every component is operating with a large safety factor. Second, the choke in the high-voltage section is operating in a resonant circuit in conjunction with C1. These two features are what make this supply different from most, and will be discussed below.

The Resonant Choke

Good voltage regulation is a prime requisite in a power supply to be used with linear amplifiers. For developing this good regulation, it is difficult to equal the choke-input filter circuit. Furthermore, the choke-input filter has other advantages. It tends to limit the peak current in the rectifier diodes, an important consideration when using high-voltage silicon rectifiers. And it allows the transformer to operate cooler than it would in a capacitor-input type circuit.

But there is one big problem when using a choke-input circuit in a medium-or high-voltage supply. That problem is that the bleed current necessary is usually quite high, unless you have lots of money to spend on high-inductance, high-current chokes. This means that bleeder resistors with a high wattage rating are necessary, and these cost money. Furth-

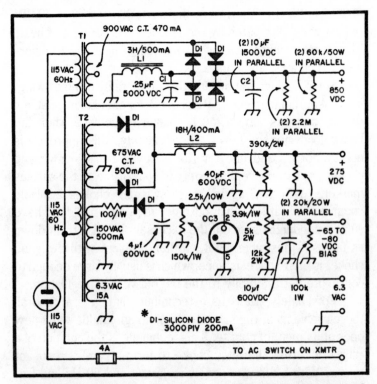

Fig. 4-12. Schematic of 850V and 275V supply with −65V bias and 6.3V AC filament sources.

ermore, and more important from a reliability standpoint, the large quantities of heat being generated by the bleeders tend to raise the operating temperature of the supply, and this in turn will reduce the life of the components.

For example, let's take a look at an 850V supply. The formula given in the handbooks for bleed current is (assuming a full-wave rectifier and a 60 Hz line frequency):

$$I_b = \frac{E_{DC}}{L}$$

where

I_b = bleed current in milliamperes
E_{DC} = DC output voltage, and
L = choke inductance in henries

With a 10H choke the necessary bleed current would be

277

850/10 = 85 mA. This means there is 0.085 × 850 = 72.3W of heat being dissipated by the bleeder. That is a lot of heat! This constant current flow also tends to heat the transformers, rectifiers, and choke.

Some designs get around this problem by using a swinging choke; that is, a choke which has a lot of inductance at low (bleed) currents, and a much smaller inductance when the heavy load current flows. Unfortunately, this approach is usually not a good one for the hobbyist because swinging chokes are expensive, and not readily available on the surplus market. And, then again, most swinging chokes available commercially (at less than outrageous prices) still don't have sufficient inductance to cut the necessary bleed current down as far as is desirable. Still another problem with swinging chokes is their DC resistance, which is usually large enough to contribute detrimentally to the overall voltage regulation.

Fortunately, there is a technique that gets around all these problems at the same time, although it adds some new ones of its own, of course. This technique is the addition of a capacitor across the filter choke to form a parallel resonant circuit. With a full-wave rectifier circuit and 60 Hz line frequency, we resonate the choke at 120 Hz, the first and most important ripple component frequency. The idea is to use a choke with a low-inductance, high-current rating. This will guarantee a low DC resistance. Then the capacitor is properly chosen to resonate the choke. The lower limit on the choke inductance is reached when the capacitor value required begins to approach the value of the output capacitor (capacitor C2 of Fig. 4-12). Usually a choke of about 2H is used, although other values are fine, depending on what you have on hand. The required bleed current with the resonant filter then is:

$$I_b = \frac{E_{DC} \times R1}{(852) \, L^2}$$

where

E_{DC} = DC output voltage,
R1 = Choke resistance at 120 Hz,
L = Inductance of choke in henries, and
I_b = bleed current in milliamperes

Now, looking at this equation, it would seem a simple enough matter to measure the resistance of the choke and then, knowing the choke's inductance and the DC output voltage, to calculate the required bleed current. But here's the hooker (remember we promised you new problems): the value of R1 can be measured, but it isn't simply a matter of putting an ohmmeter across the choke. This is because the resistance of the choke at 120 Hz won't be the same as the DC resistance, the resistance that would be measured by the ohmmeter. Furthermore, the value of the choke's inductance tends to change, depending on the current through the choke. This variance makes it difficult, if not impossible, to calculate the best value of capacitance to be used in parallel with the choke.

Once again, though, there is an answer to the problem. This time it involves what engineers call an iterative process (about the same thing hobbyists call cut-and-try). Having selected the choke you will use, you want to know the proper bleed current and the proper size of capacitor to be used in parallel with the choke. Simply follow this four step method and you will come out with the right values:

1. Take a guess at what I_b should be, say 20 mA for each 500V of DC output voltage. Then calculate the size the bleeder resistor, R_b. From Ohm's law: $R_b = E_{DC}/I_b$. For E_{DC} use 0.9 times the total transformer RMS secondary voltage if using a bridge rectifier or half this amount if using a conventional full-wave circuit.

2. Now haywire the circuit together, using R_b as calculated above. With reduced voltage on the primary of the plate transformer, say 40V, substitute different values of C1, finally selecting that value that gives the *least* DC output voltage. With a 2H choke this will be about 0.5 μF. The right value of C1 will probably be far from critical. Be sure to keep the primary voltage constant.

3. Now disconnect the bleeder resistor and measure the output voltage (continue using reduced primary

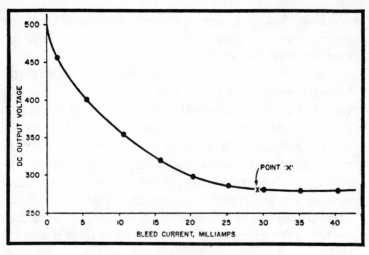

Fig. 4-13. Typical shape of curve that will result from the technique given in step 3 of text. Point X gives minimum allowable bleeder current.

voltage). Then gradually load the supply, measuring the output voltage and current at several points as the loading is increased. These points can then be used to plot a curve like that in Fig. 4-13. A good way to provide this variable load is by varying the bias on a spare transmitting tube hooked up as in Fig. 4-14. The plate dissipation rating of the tube may be exceeded as each reading is made, but if the supply is turned off between readings while the bias is reset, the tube will not be damaged.

4. The lowest allowable bleed current is that corresponding to point X on the curve in Fig. 4-13, that is, the point where the curve flattens out. Add about 30% safety factor to this value to get the proper bleed current, I_b. If the value obtained is very much different from that used in step 1, use this new value and go back to step 1, repeating the procedure. You will probably find it unnecessary to change C1 if you do repeat the steps. And you shouldn't have to repeat the procedure more than once.

All this may sound quite involved and complicated, but it takes much less time to do than to tell about it. Be sure to observe sensible safety measures when performing your measurements. When substituting different capacitors for C1, the working voltage rating can be any value greater than the output voltage, but don't use electrolytics. After the final selection is made be sure to use a high quality capacitor for this component, since the strain on it is great. The working voltage rating should be at least twice the DC output voltage.

The above process has assumed that you have already chosen the choke you will use beforehand. If a choice between two or more chokes is available, these facts should be kept in mind. First, it is desirable that the choke have as low a DC resistance as possible, in order to help provide good static voltage regulation. Second, it is desirable that the bleed current be as low as possible, and from equation 2 it is obvious the higher L is and the lower R1 is, the lower the necessary bleed current is. Thus, the ratio $L^2/R1$ should be as high as possible. If necessary, two chokes can be compared by measuring their characteristics at 120 Hz on an inductance bridge. If a bridge is not available, the only recourse is to use the choke with the higher ratio of rated inductance to rated DC resistance. This will usually provide the proper choice between two chokes.

The reduction in bleed current that can be achieved by this method of parallel resonant filters is truly amazing. In this

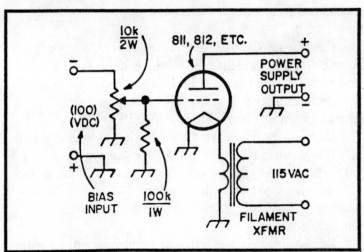

Fig. 4-14. Circuit that shows a simple approach to providing a variable resistor for testing a power supply.

supply, the bleed current in the high-voltage section is only 28 mA, and this with only a 3H choke. And with a choke this small, it is a simple enough matter to find one with a very low DC resistance.

So now we have licked the only big problem in a choke-input design. Let's look now at this business of conservative design.

Reliability

To provide reliability in an electronic design, each and every component must be operating well within its ratings. This means using components that have higher ratings than might seem necessary at first thought. Why do you suppose the military surplus components you buy are larger and heftier than their commercial counterpart? Because the millitary *must* have reliability and this requires components with a fat safety factor. (This helps make military surplus today still a big bargain in electronic components, too.)

Let's look at each type component individually and how it is selected for reliability:

Transformers and Chokes. These are fairly simple, it's just a matter of using components with a higher current rating than the minimum necessary. In this supply all the chokes and transformers have a current rating 50% higher than normal.

Rectifiers. Choose units with at least 50% higher peak inverse voltage rating than the minimum necessary. The current rating isn't quite as critical, but a hefty rating in this department is a good idea, too. This supply uses some beautiful military surplus units rated at 3000 PIV, 200 mA..

Capacitors. If you want reliability it is absolutely necessary that you use oil capacitors in your supply. Electrolytics just don't have very long lives. Follow the lead of the military and use oil capacitors with at least a 50% safety factor in working voltage rating. Oils are available in surplus at very reasonable prices.

Resistors. Heat and high temperatures are major gremlins in electronic equipment. In resistors, high operating temperatures limit component life severely. Consequently, resistors should have at least three times more wattage rating than the amount of heat they will be called upon to dissipate. This will reduce their operating temperatures considerably, and greatly increase their life. Also, note that in this supply two bleeders are used in parallel, rather than a single one. This is so that if one fails in operation, there will still be some bleed to keep the output voltage from increasing sharply and causing components in the external circuitry to fail. Also, it allows the bleeder to be divided into two parts. The two resistors are physically as far apart as possible, further reducing the operating temperature. Note that small resistors are used directly across the terminals of all filter capacitors. This is a safety precaution.

Fusing. To protect the rectifiers and other components a fuse with medium-blow characteristics should be used, with a current rating that is only slightly greater than that required for operation under normal conditions. Start with a small fuse and if it fails under the normal load, use the next higher current rating.

Chokes. Note that in the high-voltage section, Fig. 4-12, the choke and parallel capacitor are in the negative lead, between the rectifiers and ground. This is to avoid having the high voltage on the choke, reducing the chance of insulation breakdown in this component.

Chapter 5
DC-TO-AC Inverters

Let's get it straight—an inverter changes DC to some value of AC. There seems to be some confusions about just what inverters, converters, and DC-to-DC converters are and do. When talking power supplies, converters change AC to some value of DC. And DC-to-DC converters step up or step down a DC voltage, usually involving an inverter and a rectifier, but not always. Inverters will probably find there way into your automobile, as the car's electrical system is 12V DC and most equipment requires 115V AC. Here's two inverters that may help you out of a tight spot sometime.

115V 60 Hz SINE WAVE INVERTER

How would you like to have a 115V 60 Hz outlet in your car? It would be extremely convenient for use with that receiver, transmitter, or tape recorder having only a built-in AC power supply. For around $30 you can buy a little 100W solid-state inverter using power switching transistors; but wait until you connect your nice store-bought power inverter to a device having a high gain audio amplifier in it. I'm afraid you will be in for quite a shock. The sharp-cornered square wave output, often with accompanying sharp little spikes, all too frequently seeps through to your low level, high gain audio stages, producing the most annoying cacophony of sounds you have heard in a long time. The simple square wave output power inverters are great for razors, electric drills, soldering irons, and even some electronic gear, but for more demanding requirements you need an inverter with a good old sine wave output.

The expensive and number of parts required to build a sine wave power inverter are not much greater than for a noisy square wave switching-type supply. Just an additional three or four small transistors, a dozen or so resistors and capacitors, two or three pots, and a small driver transformer are all the extra parts that are needed. A glance at Fig. 5-1 will suffice to show that the basic idea of a sine wave power inverter is exceedingly simple. The block diagram reminds one of a simple 80 meter CW transmitter, and this comparison is quite valid. Basically only the frequency is different; 60 Hz instead of 3.5 MHz. And generating 100W of sine wave AC is considerably simpler at 500,000 meters (60 Hz) than at 80 meters (3.5 MHz). A typical circuit is shown in Fig. 5-2. It is not necessary to use this exact circuit; anything that you can easily devise will be fine. The essentials are a sine wave oscillator, a buffer amplifier, and a power amplifier. A few comments about the requirements on each are in order.

The Oscillator

The oscillator can be any type, but a good choice for practical reasons is one using RC combinations. Even an oscillator using an inductor-capacitor combination such as the Hartley or Colpitts can be used at 60 Hz, but the problem of varying the frequency in order to adjust it to 60 Hz is made much, much easier by using an oscillator whose frequency is determined by an RC network, since the frequency can then

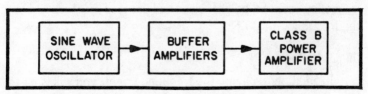

Fig. 5-1. Block diagram of the power inverter with a sine wave output.

be adjusted by merely setting a potentiometer. In Fig. 5-2 the 10K potentiometer in the oscillator section enables one to adjust the frequency. Make sure your oscillator does produce a sine wave. Some RC oscillator circuits are designed to produce square waves and others a sawtooth output. Don't use these, or you will end up with an inverter that is as good a hash producer as the simple power-switching type. The 3.9K potentiometer in the oscillator section of the circuit shown in Fig. 5-2 controls the amount of gain in the feedback loop. Too much gain will result in severe distortion of the sine wave.

The Oscillator Frequency

Incidentally, there may be times when you will want to run your inverter at other than 60 Hz. In one instance it was necessary to use an inverter to power a reel-type tape deck operating in a car. In order to run through a 60 minute reel containing voice material during a drive which normally required 50 minutes, the frequency of the inverter was increased to about 70 Hz, thus speeding up the tape deck motor. This procedure would certainly not be acceptable with music, but you soon adapt to the slightly higher pitch of voices. This approach would be a natural for varying the speed of a code practice tape. You should, however, carefully check that your equipment, in particular its motors and transformers, can be operated without overheating before subjecting it to use for extended periods of time at frequencies other than 60 Hz.

How can you set your oscillator to 60 Hz? Any of the standard ways of measuring an audio frequency can be used. You might compare your oscillator's frequency with the commercial line voltage frequency either by observing Lissajous figures on an oscilloscope or by listening to and setting the beat frequency to zero. Or if you have access to a digital frequency counter, go first class and use it to measure and set the frequency to 60 Hz.

Output Voltage

In addition to having the frequency right, you will want to have the output voltage reasonably near the desired range of

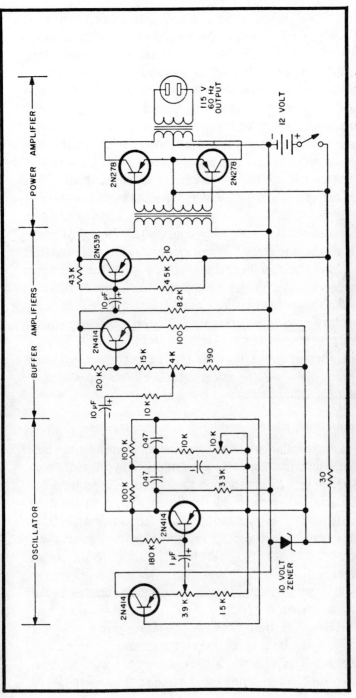

Fig. 5-2. Schematic of the sine wave inverter. The frequency can be varied by the 10K pot in the oscillator section. The transformers in the output stage can be any filament transformers with adequate power ratings.

115V to 120V. In the circuit of Fig. 5-2, the output voltage is determined by the setting of the 4K potentiometer between the oscillator and the buffer amplifier stages. Of course this pot is nothing more than what we would normally call a volume control, and the following amplifier stages are merely conventional audio amplifiers, operating in our case at 60 Hz. If you want a deluxe setup for continuously monitoring the voltage output of your inverter, you can build in a standard AC voltmeter. A simpler and less expensive approach which will enable you to monitor the output voltage quite adequately is illustrated in Fig. 5-3. A simple voltage-divider resistor network is used in connection with a small neon lamp to indicate the output voltage. The neon lamps indicated are the type requiring an external resistor for operating on 115V. Resistors R1 and R2 are selected by trial and error so that the first neon lamp ignites at 110V, and R3 and R4 are chosen so that the second neon starts at 120V. Then by setting the potentiometer controlling the voltage output so that the first neon lamp is on and the second is off, you are assured that the voltage output is in the range 110V to 120V. Depending on the type of neon lamp used, suitable values for R1, R2, R3 and R4 will lie in the range 30K to 150K.

The Buffer and Power Amplifier

The particular circuit and transistors chosen for the buffer amplifiers is completely noncritical. Any sort of audio amplifier with sufficient oomph to drive the power amplifier will be quite satisfactory. Just remember that you will need good bass response since the operating frequency is 60 Hz. In other words, don't scrimp on the size of the coupling capacitors. The same general remarks apply to the power amplifier.

The input and output transformers need not be regular audio transformers. In fact, you will be better off using ordinary 60 Hz power transformers. The driver transformer, for example, could be a small power transformer having three filament windings. One winding could be used for the primary, and the other two in series could make up the center-tapped

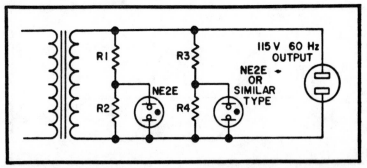

Fig. 5-3. Schematic of a simple circuit to monitor the output of the sine wave inverter.

secondary. A center-tapped filment transformer would be quite suitable for the output transformer. With a 12V DC supply, one would expect the power amplifier transistors to produce an AC voltage in the range of 7V to 9V peak value or 5V to 6.5V RMS. Thus a 12V center-tapped filment transformer with its 115V winding used as the secondary or output should be quite satisfactory. The size or current rating of the filment transformer is determined by the amount of power you will want to handle. A 5A, 12V transformer will be quite adequate for powers up to 50W or 60W. The possibility of paralleling transformers in order to take care of higher powers should not be overlooked.

The choice of transistor type to be used in the power amplifier is determined largely by how much power you need to develop, and by what transistors you may have available. Adequate cooling of the 2N278s used was insured by immersing them and their heat sinks in a large volume of antifreeze. The five leads were brought out through insulated feed-throughs soldered into the wall of the anti-freeze container, making a splash and spill proof assembly.

Efficiency

The power amplifier stage operates class B, or nearly so. The theoretical efficiency for class B amplifiers is 78.5% with practical amplifiers achieving 60% to 70% efficiency. Assuming that your power amplifier turns out to be 65% efficient and that you are delivering 100W to a 115V 60 Hz load, how many

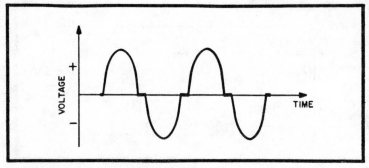

Fig. 5-4. Output waveform of the power inverter.

amps will your 12V battery have to deliver to the power inverter? Working through the arithmetic involved indicates that the load on the battery will be 14A to 15A.

Waveforms

The actual waveform obtained from this inverter departs slightly from a true sine wave. As can be seen in Fig. 5-4, each power transistor of the power amplifier has an operating angle slightly less than 180°, that is, class C type operation is being approached. A small shift in bias would bring the amplifier back to true class B, but since no noise pickup is observed, the additional effort to change the bias did not seem worthwhile.

A Caution

Transistors are quite sensitive to temperature. This fact was discovered when the power inverter, in its original form, would not function on cold winter days until the interior of the car warmed up. The oscillator would simply fail to take off because the gain in the feedback loop was too low. This sad situation was corrected by making the 3.9K potentiometer accessible so that it could be manually reset to increase the gain in the feedback loop. After the temperature rose, the gain was then lowered to prevent distortion of the waveform.

If you are looking for a simple way to run gear having built-in 60 Hz supplies in your car, this sine wave power inverter may well be your answer.

110V 600Hz INVERTER

The following data was obtained experimentally:

1. Ferrite material used in TV yokes and flyback tranformers is the same stuff they use in commercial toroids.

2. Single flyback core (larger size) or a single yoke core is about right size for 50W output operating around 400 to 600 Hz.

3. Sections can be stacked (see Fig. 5-5) for higher power.

4. Ferrite cores can be reduced to powder, mixed with epoxy cement and used to glue sections together, fill the cavities, or even to mold your own cores.

5. About one turn per volt is right for a 150W transformer, two turns per volt for a 300W job.

6. You do not need a high voltage secondary winding if you use circuit Fig. 5-6; this will reduce your coil winding to probably less than ten turns.

7. There is nothing wrong with cheap 40W transistors. If you match them, they can be paralleled, each pair producing about 75W to 100W in a 12V system.

The power supply built according to Fig. 5-7 produces 100W at 110V 600 Hz. It is using nine yoke core sections cemented together as explained in item 4 above to form a toroid. A ratio of 1.5 turns per volt was used. The primary has 18 by 18 turns of AWG #14 enameled wire. The feedback

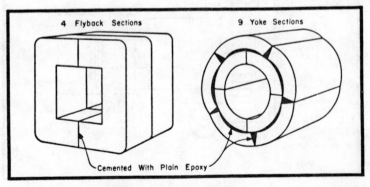

Fig. 5-5. Possible iron-core sections for an inverter.

winding has 12 by 12 turns of AWG #24 enameled wire, and the 110V secondary is 165 turns of AWG #20. Efficency is close to 80% to 85%

The circuit is a standard grounded emitter type. It uses four 40W transistors mentioned above. It is suggested that resistors R1 and R2 are made variable and an ammeter inserted in series with the feedback centertap (it reads 1A) during preliminary testing. A good heat sink is essential.

The only other part requiring some explanation is Rx. It is 1 foot of bare copper wire, coiled around a mica support, and running very hot—over 100°C—at full power (wire size will depend on the power you are running). It does three things:

1. Tends to equalize currents through four transistors.
2. Acts as a fuse.
3. Permits easy matching of the transistors, which is done by measuring voltage drop across each Rx resistor. Use test probes of VOM in the 2.5 mA position; all you need is a relative indication.

The 400—600 Hz output can be transformed to a higher voltage using separate transformer, or can be rectified using voltage doubling or quadrupling, which is very easy at these frequencies. The cost of 400 Hz transformers is quite low, and they are several times smaller than their 60 Hz counterparts.

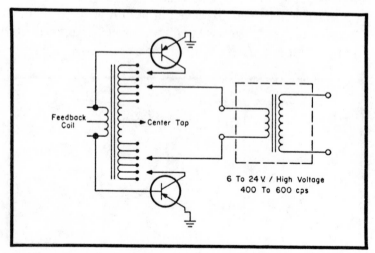

Fig. 5-6. Diagram of transformer hookup for the 110V inverter.

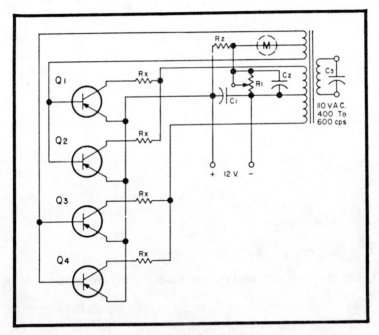

Fig. 5-7. Schematic of the 110V 600 Hz inverter.

This is course permits construction of about a 200W supply for about ten dollars, depending mostly whether you have a friend in a TV repair shop.

Parts List

Q1, 2, 3, 4 ....40W transistors
R1.......................................110Ω 5W in series with 10Ω pot
R2...150Ω 10W
RX...Equalizing resistors, see text.
C1 ....1000 μF 25V
C2 ....50 μF 25V.
C3...0.01 μF 1200V mica.
M..........................ammeter (see text) (needed during testing)
Toroid transformer constructed as in Fig. 5-6 (9 sections TV yoke cemented together).

Chapter 6
AC-To-DC Converters

As mentioned in the previous chapter, an AC-to-DC converter changes some value of AC to DC. Simply put, an AC-to-DC converter is an unregulated low-voltage supply. Battery eliminators are examples of converters. Most battery chargers also fall in this class. This chapter deals with two such circuits. There are very simple circuits yet may be useful to the low-budgeted hobbyist.

117V AC TO 24V DC CONVERTER

There is really no need for you to be passing up those 24V to 28V surplus units. For less than $20.00 you can build a power supply that will provide 24V to 28V DC at 10A.

The circuit (Fig. 6-1) is a straightforward DC power supply circuit employing four silicon rectifiers in a bridge circuit and suitable high-capacitance low-voltage capacitor for filtering.

The power supply can be built on any practical size chassis that will accommodate the components to be used. Construction is noncritical as to parts placement. A 7- by 9- by 2-inch aluminum chassis can be mounted to a standard rack panel by means of support brackets. Rack panel mounting is of course optional.

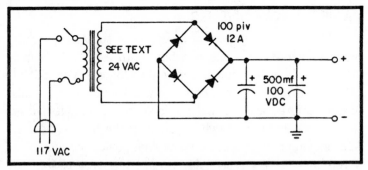

Fig. 6-1. Schematic of 117V AC to 24V DC converter.

When wiring the unit it is suggested that the builder use AWG #12 or #14 solid wire. This is the same type of wire as generally used in your house wiring. About a three to four foot length should be sufficient.

Most of the surplus stores can supply a power transformer that will do the job very nicely. Price? The range is from $5.00 to $10.00 each depending on condition and current carrying capacity.

Of course you can also obtain your 24V AC by wiring the secondaries of two 12V AC filament transformers in series (or the secondaries of four 6.3V AC filament transformers). The primary windings would be paralleled, and a check with an AC voltmeter will tell you whether you have the secondaries connected properly. Be sure to check the secondaries with an AC voltmeter to insure that your connections are series adding and not cancelling.

The rectifiers (diodes) used in the original unit are the 12A 70V RMS (100 PIV) silicon power diode studs. Be sure to use mica (or equivalent) insulating washers when mounting the diodes to the chassis.

Metering of the current drain and output voltage is optional. A DC ammeter that wasn't in use was installed on the front panel. Since a DC voltmeter of suitable range was not available at the time pin jacks were provided to check the voltage whenever desired.

A heavy duty 24V DC supply such as this will most likely be more than ample to fulfill any needs of the average experimenter or amateur radio operator.

16V OR 30V 20A CONVERTER

When an inexpensive variable power supply for mobile gear was needed this one was built with excellent results. Commercially built supplies were either too costly or did not have the current requirements needed. Construction of this supply centered around two ideas; keep it as simple and as inexpensive as possible while obtaining the highest power. After looking around a little, some high quality surplus parts were obtained very reasonably.

Because of the very nature of surplus—here today, gone tomorrow—it is doubtful if the same parts used can be easily obtained. They are included in the parts list merely as a guide. Some alternatives have been presented here to save trouble for anyone wanting to build a similar unit.

Basically, the circuit consists of a high-current, low-voltage transformer with solid-state rectifiers and an inductive-capacitance L-filter. Several refinements were added later to give the circuit shown in Fig. 6-2. A switch on the front panel (S1) provides up to 16V or up to 30V at 20A through a relay which changes the windings on the secondary of the power transformer. Meters showing voltage and current are provided and the voltage is varied by means of a small variable transformer mounted on the front panel.

The main component of this type of power supply is, of course, the power transformer. The secondary consists of four 9V windings at 10A each. Filter chokes T2 and T3 are both 10A.

The diodes, meters, switches, and output connectors can be found almost anywhere. The 0 to 30A meter is for a car. The variable transformer is a small 2A unit which came out of a piece of fire-control equipment and varies the voltage on the primary of the power transformer. Two 1000 μF, 25V capacitors were used in series at first, hoping they would work, but it was found that the ripple was too high with T2 and T3 at 0.037H. Several of these capacitors were on hand, so they were wired in series-parallel until there were sixteen, totalling 8000 μF. Not very professional, but quite effective. Two 8000 μF units at 55V each would work even better in parallel. If you use odd values in series be sure to shunt them with a 3.3K, ½W resistor to equalize the voltage drop across the capacitors.

Construction is straightforward and wiring noncritical. The front panel should be attached to the chassis first and then the switch and pilot light holes drilled. Locate the rest of the parts on the chassis and locate their positions, then drill the holes. If a variable transformer with an external wiper arm is used, make sure there is clearance with the arm in all positions. Control wiring-switch contacts, variable transformer winding leads, and pilot light terminals are made with regular solid hookup wire, but use heavy wire for all currrent carrying leads. Use stranded AWG #14 wire for easier handling. Manipulating the power transformer's solid leads proved to be

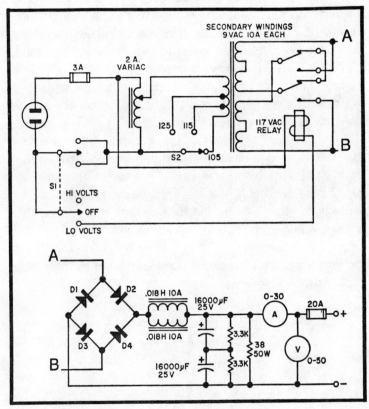

Fig. 6-2. Schematic of 16V or 30V 20A converter.

297

very difficult. If you get a transformer like the one used here, any heavy duty relay can be used to switch the secondaries. A pilot light enables you to tell when you are in the low-range position. It is connected across K1 and lights when the relay is energized. The diodes run cool mounted to the chassis with no heat sink.

The diodes are rated at 20A each. A circuit breaker should be included in series with the output to prevent damage to the diodes and other components because the fuse in the primary of the transformer will not act fast enough.

With the power supply in operation, regulation is poor because of the charge of the capacitors. However, the variable transformer can be adjusted to compensate for different loads. On-off switch S1 is a DPDT type with a center-off position. In the low-range position, with the toggle pointing downward, the relay kicks in to place the transformer windings in parallel. In the high-range position, the toggle is up and the relay is not energized. Because of the instant-on feature, the switch should be wired this way to prevent accidentally placing a higher voltage on units under test. To turn on a piece of equipment, the operator will instinctively push the switch down to get a low-range voltage.

This supply has been used for servicing transistor radios and running mobile equipment and unconverted 28V surplus gear with no trouble. Surplus parts are used throughout except for the two pilot lights. And from the use I get out of it, the parts have proved to be the best deals I ever made. This under-twenty-dollar 20A supply would probably sell for well over a hundred dollars retail! So, if you are short on cash and high on amps, try this one. It's a shame to let all those high quality parts go to waste.

Chapter 7
DC-To-DC Converters

With the complexity of today is electronic equipment you can seldom get by with just one supply voltage. Interface equipment also requires furnishing several sink and source voltage. There are many methods of generating the required potentials, but one quick and easy way is to employ a DC-to-DC converter.

In reality most DC-to-DC converters are two circuits in one, an inverter and a converter. The basic inverter employs a DC supply voltage to generate an AC signal, usually at a much higher frequency than 60 Hz. To use a 60 Hz oscillator in the inverter circuit would require large and bulky components, especially if any amount of current was desired. Therefore many inverters operate above 1 kHz.

Most inverters are driven very hard, resulting in a square wave output. The output is coupled through a transformer to a rectifier circuit, where it is converted in to DC, then filtered. A portion of the square wave output is coupled back to the oscillator to maintain oscillations. All oscillators require feedback to keep going.

Nearly all of the circuits described in this chapter are of this type. One uses an IC timer to generate an AC signal for rectification, and another, a step-down voltage converter, uses a zener voltage divider.

POLARITY CHANGER

A problem often arises with some IC or transistor circuits when just a modest amount of negative voltage is needed to bias a switching circuit or to power an IC stage that works best when it is powered from equal positive and negative supply potentials. For home equipment, this situation is easily resolved by constructing a power supply which provides the necessary positive and negative outputs. For battery-operated equipment, an extra battery can always be provided, but this adds to bulk, eventual cost and perhaps erratic operation at some time since the separate batteries used for positive and negative supply voltages will not age at the same rate.

The greatest problem is usually with adapting such circuits to mobile operation where only +12V is available. Either a battery for a negative supply has to be used, the circuit is modified for single supply operation, or, as is usually the case, the idea is given up of using a particular circuit for mobile operation.

Here a number of ways are described by which a negative supply voltage can be generated when only a positive source is available. Those who still remember vacuum tubes may also remember that this was also a problem then for mobile operation when tubes required a negative bias voltage. The usual solution was another winding on a vibrator power supply transformer or on a dynamotor. But, some manufacturers even then did try more ingenious approaches, and they are the basis of the circuits that are used these days to generate that needed negative supply potential.

Basically, there are two approaches that can be used to generate a negative ground potential when only a positive to ground potential is available. One is to build an oscillator, operating anywhere from AF to RF, which has a transformer output isolated from ground. The transformer output is treated like the secondary of an AC transformer (rectified and filtered) to produce a negative voltage. In tube days, RF oscillators were popular because high voltages for bias purposes could be developed across their tuned circuits. The other basic approach is to use a switching circuit to charge a capacitor with a positive supply. But, between charging intervals, additional circuitry literally lifts the capacitor up and turns it around with respect to ground, and it supplies a negative potential. So, the capacitor is alternately charged and electronically flipped around and discharged.

The following oscillator circuits are not the simplest form of positive to negative supply generators possible. Still simpler circuits are possible but they require special transformers, whereas the two circuits shown need only simple, readily available parts.

The circuit of Fig. 7-1A is basically just an audio oscillator with a transformer coupled output. It can be constructed using a wide variety of PNP general purpose transistors and with transformers having varying turns ratios so that output voltages of from 6V to 20V can be obtained. The secondary voltage is then rectified using a standard bridge rectifier circuit and a 500 μF filter capacitor. Not too much filtering is required since the oscillator produces a fairly good sine wave output. One may have to connect miniature transistor output transformers back to back to obtain a desired output voltage, depending on which types are available locally. For instance, one might use a 400Ω CT to 8Ω transformer back to back with an 8Ω to 10K transformer to get a voltage stepup. Some experimenting and bread-boarding is necessary, but just about any desired voltage can be generated by using different transformer ratios and types of rectifier circuits (half wave, full wave, or bridge). Generally, using low power transistors and the usual 99¢ 300 mW transistor transformers, output loads of up to 10 mA can be accommodated.

Figure 7-1B is a standard multivibrator circuit with component values chosen so it operates at about 7 kHz. The idea is basically the same as in Fig. 7-1A, but this circuit produces a square wave output and, depending upon the power transistor used, output currents of an ampere or more can be generated at various voltage levels. Some 24V CT filament transformers will work well in this circuit. Otherwise, standard transistor output transformers have to be used either singly or back to back to get the desired step up in voltage. Transistor output

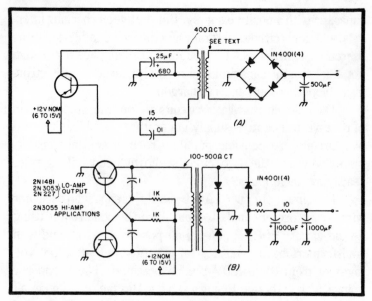

Fig. 7-1. Two forms of oscillator circuits which can be used to form a negative voltage supply. Output voltages and currents are available over a wide range depending on the transistors and transformers used as explained in the text.

transformers or interstage transformers rated as 1W such as Stancor TA-4 or Allied 6T8HF, will do nicely with any of the low/medium power transistors mentioned in Fig. 7-1B to produce outputs in the 12V to 15V range at up to 100 mA load. The square wave output of the circuit requires that a bit more elaborate filtering be done, as shown in Fig. 7-1B, to get a smooth DC output. Also, square wave generators will have rich harmonic outputs. The 7 kHz operating frequency for the multivibrator is a compromise between an operating frequency that is above the audio range of most audio communications circuits with which it is likely to be used and yet within the efficient frequency operating range of most inexpensive transformers. Nonetheless, if the supply is to be used with RF circuitry, careful power supply bypassing to the multivibrator and even shielding in extreme cases may be necessary to prevent hash in the RF circuits.

The circuits of Fig. 7-1 should satisfy most needs and can be built reasonably inexpensively. However, in these days of ICs, it would not be fair to omit some more sophisticated approaches to the negative supply voltage problem. Both of the following circuits appeared originally in *Electronics* and provide negative voltage generation along with regulation of the output voltage.

Figure 7-2 uses the ubiquitous 555 IC timer. The 555 is operated as a free-running pulse generator. The pulse output frequency and pulse width are basically set by the RC components between pins 7, 6, 2, and 1; however, a control voltage applied to pin 5 can vary the pulse repetition rate. Output is taken from pin 3, where first a capacitor isolates the output from a ground reference and then the output is rectified and filtered. The output voltage level can be set anywhere from 0 to −10V by the potentiometer in the base of the 2N2222. This potentiometer, for any given setting compares the output voltage to the supply voltage. If the output votlage decreases, the 2N2222 is turned further on, causing the control voltage on pin 5 to come closer to ground, hence to increase the pulse

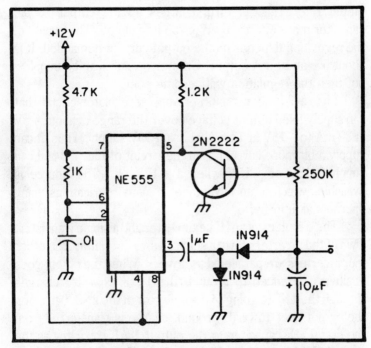

Fig. 7-2. The inexpensive 555 timer IC is the heart of this negative supply which provides a variable output up to − 10V. Note that the 1 μF capacitor by pin 3 is not an electrolytic type.

303

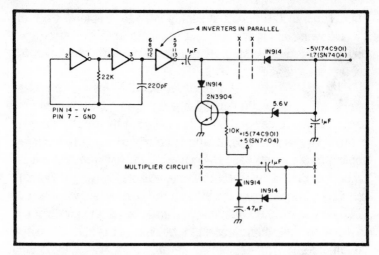

Fig. 7-3. Negative voltage generation circuit using a hex inverter IC. If the multiplier circuit is inserted between the Xs, the basic output voltage will be the same as the positive supply voltage. The zener in the base of the 2N3904 must always be rated at 0.6V to 1V more than the final output voltage.

output rate. This, in turn, charges up the output 10 μF capacitor more often and brings up its voltage. The quality of the regulation depends on the output current demanded. It is about 5% at 10 mA output. Greater outputs can be supplied but then the regulation will become poor.

Figure 7-3 shows another supply/regulator circuit that can supply fixed output voltages over the range of about −5V at 30 mA to −15V at 13 mA from a +15V supply. Thus, it can supply a bit more current than the circuit of Fig. 7-2 and can supply −12V or −15V when a +12V or +15V source is available, which is an advantage in many applications using operational amplifiers.

The circuit uses an IC hexinverter and a single transistor. Two of the inverters form a square wave oscillator. The other four inverters are paralleled as drivers. When the output goes positive referenced to ground, the 1 μF capacitor charges through the 1N914 going to the collector of the 2N3904. When the output goes towards ground, charge is transferred from the first 1 μF capacitor to the output 1 μF capacitor as the second 1N914 conducts. The output 1 μF capacitor now has a voltage across it which is negative referenced to ground. This transfer of charge between capacitors continues until the voltage buildup across the output 1 μF capacitor breaks down the zener diode. Then, the 2N3904 is turned off and the transfer of charge stops until the load current drain again causes the output voltage to fall below the zener voltage. Thus, the circuit is self-regulatory.

The circuit as shown will produce only -5V output from a $+15$V supply. This can be increased by using the alternative voltage multiplying circuit shown in Fig. 7-3 to produce -15V. Additionally -12V can be produced from a $+12$V supply with the save circuit by using an approximately 13V zener. The 74C901 hex inverter operates from up to a $+15$V supply and can produce a full -15V when used with the voltage multiplying circuit. However, if only a $+5$VV source is available and -5V is needed, the basic circuit of Fig. 7-3 with the voltage multiplying circuit can be used with a simple SN7404 hex inverter. An approximate 5.6V zener will suffice for the zener reference diode in the base circuit of the 2N3904.

12V TO 24V CONVERTER

In mobile operation, it is frequently desirable to increase the battery voltage to some higher value. Although this section specifically describes a unit for converting 12V to 24V, the circuit can be used for other voltages up to about 35V with a power limitation of about 125W. The particular unit described is being used to power a transistorized transmitter which wouldn't draw sufficient current with 12V input. A smaller unit could be built to run a 12V receiver in an older 6V Volkswagen, a larger one to power some of the surplus 28V equipment on the market. The circuit is unusual in that it uses no rectifiers to change the AC to DC. In fact, it can be made to work with only three components.

When it was stated that the circuit uses no rectifiers, it may have been somewhat misleading. In actual fact, the transistors that do the oscillating also do the rectifying. To see how this can come to pass, let's look at the circuit. Referring to Fig. 7-4, assume Q1 has just been turned on. There is a current in the upper half of the primary and 13V is being developed in each half of the feedback winding. Summing voltages around the circuit, we have 13V between A and B, −1V between B and C (because of the forward conduction of the emitter diode), and 12V (the supply voltage) between C and D. This gives a total of 24V between A and D. Although there is 13V developed in the other half of the feedback winding, it is not conducted because the emitter diode of Q2 is reverse biased. When the core saturates, the current in the feedback winding reverses and turns on Q2. The relative polarities remain the same, so there is 24V DC between A and D, with the two transistors acting as a full wave rectifier. C1 is a filter capacitor and R1 is for easier starting under no load.

Transformer design is indicated in Fig. 7-4, using a core made by Arnold Engineering (5387-D4). If you wish to design for different voltages, remember that an extra turn or two should be added to the feedback winding to compensate for circuit losses and voltage drop in the emitter diode. Most of the precautions normally taken for preventing high voltage damage can be ignored since there is no high voltage present. Both windings were wound directly on the core using the spaces between one winding for the other. The toroid was then covered with one layer of electrical tape and mounted by its leads on two terminal strips.

As can be seen from the schematic (Fig. 7-4), the entire DC output current flows through the bases of the transistors. Since the base current rating of most power transistors is one third to one fifth that of the collector current, the most economical use of this circuit would be to multiply the voltage about four times. If you need 24V in a 6V car, you're in luck. If you want 24V from 12V, the circuit will be just as efficient but part of the capability of the power transistors is wasted since the collector current must be limited to twice the base current. If substantial power is required (over 30W), it is necessary to use auto radio transistors. It is not recommended that the output voltage be greater than 35V unless you're sure that the transistors can withstand the voltage. It might be a good idea to put spike suppression capacitors from the base to the emitter of each transistor in this case.

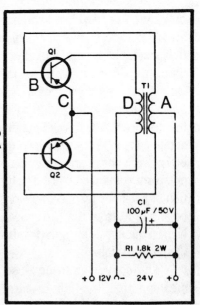

Fig. 7-4. Schematic of the 12V to 24V converter. It's good for about 1A output at 24V.

12V TO 100V CONVERTER

Often it becomes desirable to include a single tube in an otherwise transistorized system for high-input-impedance, dynamic-range, or AGC reasons. A case in point is in the RF front end of a VHF mobile receiver, otherwise all transistors. It was found to be the least trouble in this receiver to use a single 6DS4 Nuvistor as the first RF amplifier. This tube provided good AGC characteristics, freedom from cross modulation, and low noise figure.

But supplying plate voltage was a problem because the commonly available converter units were for much larger power requirements. Ordinarily, the upper frequency of amateur DC-to-DC converters is limited by the transistor cutoff frequency. With the less expensive germanium power transistors running at about one-fifth their cutoff frequency, DC-to-DC converters sing out loudly at several kilohertz. (The rule-of-thumb seems to be to run the transistors at one-fifth their cutoff frequency so as to achieve fast rise time of the square wave.)

However, for this application we are *not* constrained to use germanium power transistors, since we wish to convert only a watt or two. For switching currents of this order there are many inexpensive germanium transistors that have cutoff frequencies of several megahertz. Now, it is the core material that is the determining factor in limiting frequency, so we can almost immediately forget C-cores and tape-wound types. We are now in a good position to exploit the wonderful world of ferrite cores, which will function on up into the megahertz. An additional bonus presented to us by the ferrite manufacturers is that such cores are not only available in torroid forms but in pot-core forms which are much easier to wind. The combination of pot-core form (where one simply winds a small plastic bobbin right from the wire spool, with no shuttle needed) and high frequency operation (few turns are needed), really makes this an easy job.

The converter to be described runs at about 20 kHz, meaning that the job of filtering, after the rectifier, is simplified. Further, the 20 kHz note, due to any magnetostriction, is above audibility and shouldn't bother anyone but the family dog. The circuit is shown in Fig. 7-5; note the 1500 pF despiking capacitors (yes the despiking capacitors get smaller

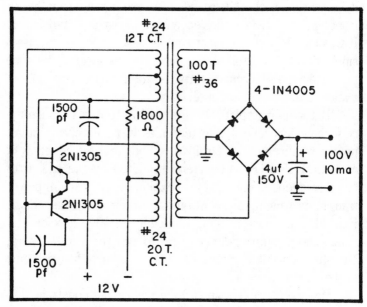

Fig. 7-5. Schematic of the 12V to 100V converter.

too as operating frequency goes up). The unit uses 2N1305s and 1N4005s for a total semiconductor cost of about $3.50. The core with bobbin was obtained locally at a surplus electronics dealer but a standard Indiana General pot-core form such as a CF214-H could be used. The local cost with bobbin was about a dollar; which is probably the better way for an individual to buy the cores as the manufacturer and his representatives have a $10 minimum order policy.

The primary windings were wound on the bobbin first although with the completely closed magnetic configuration of the pot-core construction it probably is not critical which winding goes on first. The secondary was then wound over the others and taped with tape. The windings were all hand wound which took only a few minutes. No winding care is necessary as the bobbin is far from being filled up. The pot-core halves are held compressed together by a 6-32 nylon screw and nut and the eight wires brought out the various holes in the core.

The unit runs about 60% efficient which isn't so great as DC-to-DC converters go but at this power level it is difficult (and also of small importance) to achieve more efficiency.

12V TO 350V CONVERTER

The circuit in Fig. 7-6 has been used by many. Two 6.3V filament transformers are used; one as a power transformer and one as the feedback transformer. The basic circuit is a multivibrator. One transistor is cut off while the other is conducting. The collector current of other is conducting. The collector current of the conducting transistor passes through half of the 6.3V winding of the transformer T1 similar to the operation of the standard vibrator circuit. The primary of this transformer steps the voltage up to a large value. This induced voltage is rectified by the solid-state diode bridge circuit and filtered by C2.

Feedback transformer T2 steps the high voltage AC down to a low level for drive to the bases of the transistors. Q1 is driven to cut off when Q2 is driven to saturation and vice versa. Resistors R1 and R2 set the operating bias.

The transformers used are not ideal for the job as the normal operating frequency is about 400 Hz and the core material should be square loop material. However, the circuit as described is quite practical. It may be worth while to experiment with 400 Hz transformers if you have some available.

A 1.2A filament transformer was used for feedback transformer T2 because it was readily available. Any size, smaller or larger will do. Nothing is gained using a larger one as little power is required. Stepup transformer T1 is a 6.3V, 3A filament transformer. A higher current transformer can be used, giving slightly higher voltage and more power. You are cautioned, however, to use good heat sinks on the transistors and check for excess temperature. There is little impedance in the low voltage windings to limit collector current to a safe value. Some experimenting was done with other transformers, but the 3A transformer seemed to be a safer design. It will run warm in normal operation. A 5V filament transformer will give higher output voltage but current available will be less to keep within the power ratings.

The waveform of the high voltage AC and the waveform at the transistors (collector-to-collector) should be a square wave. There should be no spike on the leading edge of the square wave as this could cause breakdown of the transistor. The purpose of the filter across the 12V supply is to eliminate this spike and to filter out line noise.

Theoretically, the supply should not operate unless the feedback winding to the transistor bases are of the proper phase, but it was found that it would work with the wrong phasing at much reduced output. Efficiency is much lower and collector current is high. Also the waveform is closer to a sine wave than a square wave. This condition forces the transistors to remain in the linear portion of their curves for a longer period during each cycle, causing them to overheat. Do not allow this condition to exist. To reverse the phase of the feedback transformer reverse the two primary leads of T2.

The purpose of the resistors in series with the primary of the feedback transformer is to limit the base current into the transistors. Also, if the output of the supply is accidently short-circuited, the feedback current drops to a very low

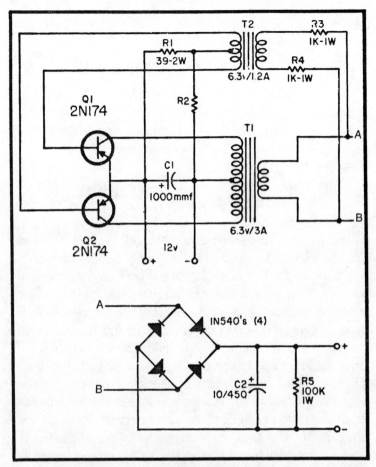

Fig. 7-6. Schematic of the 12V to 350V converter.

value, causing the circuit to cease oscillation. This provides some protection for the transistors; however, the circuit is fused as an additional precaution.

The output of the transformer is rectified by a silicon diode bridge. This eliminates heater current and gives a higher output voltage.

The circuit is designed for 12V operation; however, output starts at a low value of input voltage. The same circuit can be used for 6V input supplying 150V at about 100 mA. The listing on following page shows the output voltage versus load for 12V input using the components shown in Fig. 7-6.

Input Current at 12V	Output	Efficiency	Output Current
2.4A	400V DC	26 mA	34.9%
4.4A	350V DC	97 mA	64.2%
5.8A	300V DC7	160 mA	69.0%

Construction

The supply is built in a 3- by 5- by 7-inch box with all components mounted on the cover. All parts are inside except the transistors.

The case of the transistors is the collector terminal and must be insulated from the chassis. A fiber shoulder washer was used under the mounting nut. Sheet Mylar was used between the transistor case and the chassis, but mica or other insulating material with good heat transfer may be used. If there is danger of shorting the case of the transistors to ground where the supply is mounted in the car, a perforated cover could be used to protect them and still allow free air circulation. Use an ohmmeter to check each terminal of the transistors for shorts to the chassis before connecting wires to the terminals. By keeping all transistor circuitry above ground it is possible to use the same supply for either a negative or positive grounded battery by grounding the proper terminal.

6V TO 12V CONVERTER

The following two circuits designed to provide 12V from a 6V car system were built and tested. Both require no changes in your present electrical system other than inserting one resistor.

In the first one (Fig. 7-7) the value of resistor R1 is chosen so that a voltage drop across it is 6V with charging current to the battery held constant and reduced about 50%. This will make available 12V at terminal A of the generator. A typical case with a 30A generator will look like this: the generator is producing its normal 30A, but the voltage at A is now 12V instead of normal 8 because the field coil has 12V/3A

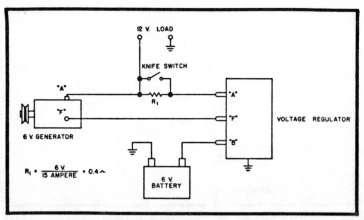

Fig. 7-7. Crude method of obtaining 12V from a 6V system. Don't look for the generator to last long.

across it instead of usual 8V/2A. The 12V output is split in two equal parts—15A is available for your 12V load (180W), and the other 15A are dropped by R1 to 6V and are used to charge the battery. Your voltage regulator MAX current contacts are adjusted to open at 15A instead of 30A, and their action will keep the generator output constant at various engine speeds.

It is recognized that this method puts 50% overload on the generator, but if your engine does not overheat and you do not drive through the desert, it seems to be perfectly safe. When 12V are not needed, the switch is closed and the system returns to normal, except that the maximum charging current is now 15A (10Ω across F and GND on your regulator will restore it to 30A if necessary).

A word of caution—open R1 will result in a burned out generator, better use a 150-200W resistor.

A second circuit (Fig. 7-8) makes use of an ordinary transistorized DC-to-DC converter. Both input and output are 6V, the output is simply connected in series with the car battery providing 12V output for the load.

No transformers are available commercially but they can be easily wound at home. No calculations are necessary—you just pick up any 100W 60 Hz transformer that has a 6V winding on it and remove all wire, counting the number of turns on the 6V winding only. Using the heaviest wire possible, wind two

313

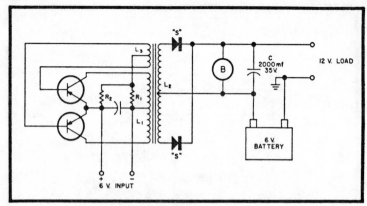

Fig. 7-8. Schematic of the 6V-to-12V converter. The circuit produces 6V out, but the output is in series with the 6V battery.

identical centertapped windings using the same number of turns *each side of centertap* as the original 6V winding had. These windings will be L1 and L2. L3 is same as L1-L2, but uses thin wire about AWG 28 to 30.

Item B is a 12V light bulb. It serves as a bleeder passing about 1A and doubles as an indicating light showing the presence of an output.

The 6V, 12A output (72W) in series with your car battery will provide 140W of 12V output to the load. Item S, are silicon stud rectifiers rated 50 PIV at 15A. Transistors are the same as you would use for a regular 6V transistorized power supply delivering the same wattage output.

The same ideas can be used to get 24V out of a 12V car electrical plant.

12V TO 6V CONVERTER

Mobile gear designed for 6V batteries is often sold at a bargain price simply because 6V electrical systems are not common. The classic method of using a dropping resistor wouldn't work in all cases. Most transceivers require almost ten times as much current while transmitting as it does when receiving. With a dropping resistor this means providing the proper current and voltage while receiving, then increasing the source to get 6V when trying to draw ten times as much current for transmitting.

The solution is to build a device that delivers a constant voltage under all conditions. This device would also have to handle currents of from 1A to 15A and yet be cheap enough to make buying 6V gear still seem attractive. A little investigation points to a solid-state regulator as probably the best answer. True, it would be as inefficient as a dropping resistor powerwise, but for purposes, simplicity and low cost are more important than power efficiency. So with these requirements in mind, the circuit shown in Fig. 7-9 was developed. It worked so well that it was thought others with 6V gear and 12V vehicles would like to try it.

Operation

When switching from transmit to receive, the current to a typical transceiver jumps from 1.5A to 12A. As this happens, the current to the base of Q1 changes from a few microamps to only a few milliamps due to the current gain of the transistors. The base current for Q1 is drawn from a point at 8.2V; voltage is produced by a zener diode and a 68Ω resistor. The zener diode maintains essentially the same voltage when the current through it changes.So with about 100 mA flowing through the diode, little happens to the 8.2V when a few milliamps are

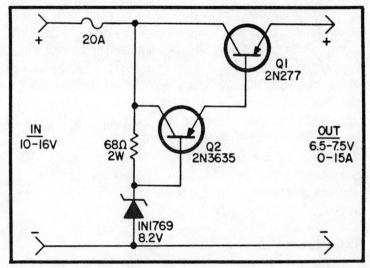

Fig. 7-9. Schematic of 12V to 6V converter.

diverted from the diode to the base of Q1. Transistors Q1 and Q2 have only current gain and no voltage gain due to the way they're connected. This means that the 8.2V of the zener appears across the load. Actually, there is a small voltage drop through each transistor, and the voltage at the output is about 6.5V.

Parts and Construction

When first beginning to collect parts for the regulator a bargain power transistor and a heat sink was purchased from a mail order surplus dealer. The rest of the parts can be scrounged from your junkbox. A 2N3635 happened to be the best transistor on hand with the right characteristic to drive the power transistor. There's a large group of cheaper transistors (2N2147, 2N4314, 2N3613) that will work as well. If you don't use a 2N3536 you'll find that the regulator parts are low cost, easily found items.

The parts you end up with and the space available will determine the exact layout. Placement of parts is not critical, but try to put the regulator in a spot where heat will not be trapped. For best heat removal, mount the regulator on a large metal surface with the heat sink fins open to the air.

The test model was mounted on the back of a transceiver. This allowed the chassis to carry away most of the excess heat. The 2N3635 and the 2W resistor normally run warm, too, so epoxied both to the underside of the heat sink. To hold the zener and serve as a strain relief for the heavy wiring, make a small fiber board card. The card made here fit nicely under the center section of the heat sink despite having some of the space taken up by the epoxied parts. From the card you should run AWG #14 stranded wire through a 20A fuse to a plug designed to fit a cigar-lighter socket. With everything except the fuse enclosed by or mounted on the heat sink, the tested regulator proved to be both compact and rugged.

Checkout and Alignment

When the regulator is completed it will require some simple tests before it's ready for operation. Carefully check all the wiring before connecting it to a 12V supply. Because of the high current involved, a wiring mistake could be dangerous. As a check on the output voltage, connect about a 10Ω 10W resistor to the output and a 12V supply (or car battery) to the input. Measure the voltage across the 10Ω resistor (should be about 7.5V). This will be the output voltage under light load. When the current drain is raised to 15A, the voltage will drop by about 1V. The drop is usually all right since most 6V gear will work from a 6V to 8V supply. If you feel that the output voltage is too high for your application, change the 8.2V zener diode to a 7.5V diode. This will drop the output voltage by 0.7V.

Comments

Once testing and alignment are completed, the regulator should be in good working order. The output voltage will stay almost constant even if the supply voltage goes as low as 10V or as high as 16V. And currents of 1A to 15A can be drawn with only a small change in output voltage. It is not recommended applying more than 16V to the regulator or drawing more than 15A through it; such conditions exceed the ratings of the power transistor. Also, don't try to draw maximum current continuously if your heat sink is not capable of dissipating at least 150W. If you don't intend to draw high current or you draw high current intermittently, a smaller heat sink can be used.

12V TO 300V CONVERTER

Recently, when confronted with the problem of operating a surplus vacuum tube receiver from the automobile supply, it was decided to use a transistorized DC-to-DC converter circuit to supply approximately 300V DC at 50 or 60 mA. After a search through a reasonable amount of literature, it was discovered that there is a serious drawback to these types of circuits. The most usual case is one in which an elaborate transformer (usually having five windings) is used in a push-pull class B arrangement with the extra windings usually for feedback. If this transformer can be purchased after three months of letter writing it is usually expensive. The other alternative is to wind your own transformer by going down to your local distributor and purchasing a transformer core, assorted enameled wire, and various insulating materials. If one is lucky enough to finally get his transformer wound, it is a feat to be proud of.

If only a standard off-the-shelf transformer could be used it would be a tremendous help. The circuit shown in Fig. 7-10 is such a device. A standard 12.6V CT filament transformer is used in reverse, driven push-pull by a pair of large power transistors, which in turn are driven by a conventional astable multivibrator. PNP transistors are used for the multivibrator, so that base current in the power transistors can be accurately controlled. The circuit configuration also provides high isolation between the oscillator and amplifier. The 2N1305s are very inexpensive switching transistors, and the entire cost of the oscillator portion is approximately $1.50. Since an inexpensive transformer is used, the cost of the unit is substantially reduced, even though extra circuitry is required to supply the drive to the power transistors. The power transistors are specified as 2N3055s but the circuit is very flexible, and any of the bargain-type NPN power transistors should work equally well, although to guarantee operation with minimum debugging it would be best to stick to the circuit components listed.

The frequency of oscillation is approximately 1500 Hz. This frequency allows good filtering with small capacitance and yet is low enough to enable a reasonable amount of power to be transferred through the transformer. A half wave voltage doubling circuit is used, but this is entirely up to the builder, depending on his particular requirements.

Layout is not at all critical but note, care should be taken that at no time any part of the high voltage circuitry be allowed to come in contact with the transistor circuitry. This would have a devastating effect on the transistor junctions.

The two power transistors can be mounted on one heat sink. Care should be taken to insure that the power transistors are insulated. Mica washers and thermal grease should be used. Before applying power, an ohmmeter should be used to insure that the power transistors are indeed insulated.

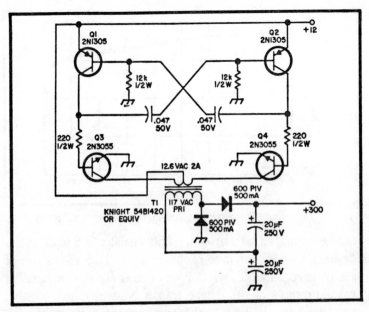

Fig. 7-10. Schematic of 12V to 300V converter.

When power is applied, a high pitched whine should be audible, which indicates that the transformer is being driven. If no whine is audible, then place a pair of conventional 2000Ω earphones from collector to collector of Q1 and Q2. The earphones should sing loudly, indicating that the astable circuitry is indeed oscillating. If these tests indicate the circuitry is operating, check for high voltage output. (Note, when this unit is delivering full load it requires approximately 4A at 12V DC). The load curve shown in Fig. 7-11 is for 12V DC, input, using component values as listed. To decrease output voltage for a particular load, resistors R3 and R4 can be increased, but it is not recommended that they be decreased to obtain higher voltages.

12V TO 210V CONVERTER

This circuit (Fig. 7-12) is a novel one. It prevents any switching transients from occurring, thereby removing this cause of transistor failure. The reason for this is the fact that the feedback transformer (T1) secondary is equally loaded through all parts of the switching cycle. Reverse voltage on

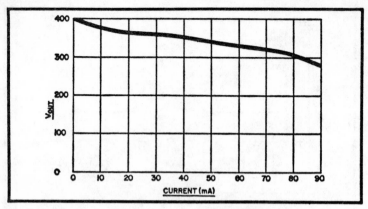

Fig. 7-11. DC output voltage versus output current for the circuit of Fig. 7-10.

the base-emitter junction of the off transistor is limited to the diode (D1 or D2) voltage drop across it. This allows use of inexpensive 2N3055s, which have emitter-to-base breakdown voltage ratings of 7V. These transistor, which are available very reasonably have collector current ratings of 15A maximum, so no larger heat sinks are required.

The main power transformer (T2) is a dual-winding 6.3V filament transformer. The secondary windings are used as the primary and vice versa.

After much experimentation, it was found that a small 115V-to-24V 250 mA unit made a good feedback transformer (T1). When connected as shown, it furnishes more than enough base current to insure driving the 2N3055s into saturation, providing low switching losses and good efficiency. Measurements indicated 315 mA of base drive per transistor. According to specifications, this is sufficient for switching 9.45A amperes collector current, assuming a minimum beta of 30.

Resistor R1 provides a small base-emitter forward bias. As the secondary DC resistance of T1 is only 8Ω and small by comparison to R1, it is not necessary to bias each transistor with a separate resistor. Imbalance is immeasurable.

Capacitor C1 helps filter out any transient spikes appearing on the 12V line. The secondary circuit is a conventional bridge rectifier and filter setup which furnishes about 210V under transmit conditions.

320

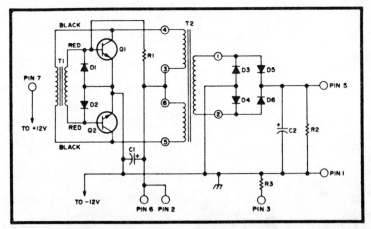

Fig. 7-12. Schematic of the 12V to 210V converter.

This same circuit can be used with regular 12V vibrator transformers, making cannibalization of an old car radio worthwhile to obtain a suitable power transformer.

Owners of earlier FM gear using vibrator supplies can transistorize them inexpensively, gaining efficiency and reliability in the process. Just remember to provide adequate transistor heat sinking if you plan on making a 100W or 200W supply, for example. The circuit will also provide an ideal receiver supply for 450 MHz FMers who want to "duplex" their surplus mobiles.

About the only difficulty which might be encountered in the construction of this supply is its failure to oscillate. Should this happen, merely reverse *one* set of leads on the feedback transformer.

Don't worry about overheating *either* transformer.

Parts List

T ....115V to 24V 250 mA (Fair Radio MW#4528)
T2......................115V to 2× 6.3V 1.2A (Fair Radio #7629809)
D1, D2 ....Silicon diode, 1A 50 PIV
Q1, Q2...2N3055 (Poly Paks)
D3, D4, D5, D6, .Silicon diode, 1A and 400 PIV or more (Poly Paks)
C1...100 μF 25V electrolytic
C2...20 μF 350V
R1...1000Ω
R2..470kΩ
R3 ....150Ω

Chapter 8
Battery Chargers

A battery charger is a power supply. In fact, the identical components, engineered into circuits that serve as battery eliminators for portable equipment, go into chargers. Very often you'll see battery charger for nicads serving double-duty as a battery eliminator. This chapter has several such chargers, for all types of batteries, even for the silver-zinc and gelled cells.

REGULATED NICAD CHARGER

To avoid damaging nicad batteries a few precautions are necessary:

1. Always utilize the full capacity of the cell. Nicads have a sort of memory action, and a unit that is habitually required to provide only half of its rated capacity will go dead at that halfway level when the whole bit is needed.
2. Don't reverse charge a nicad. Keep the charge condition on all cells in a series string at the same percentage rate. Substitution of a partially charged cell into a series string of fully charged units may ruin the weaker cell through reverse charging.

3. When charging standard nicads limit the charge current to about one-tenth the rated ampere-hour capacity. Excessive charge current causes overheating, which may result in seal rupture and venting of excess pressure. Once the seal is broken, the cell will rapidly dry out and become useless.

Figure 8-1 is representative of a universal type nicad charger circuit.

The transformer, rectifier, and filter capacitor are conventional design. The transformer itself is an 18V doorbell unit which gives a rectified DC output of 25V.

The current regulator is somewhat less conventional, as most are familiar with the emitter follower circuit in Fig. 8-2. Placing the load in the collector circuit as in Fig. 8-1 allows a measure of gain and results in better current limiting action.

In Fig. 8-1, resistor R1 is used to provide forward bias to the base of Q1, bringing that transistor into conduction. With no collector load (batteries) in the circuit, the emitter current is very low. Thus the resulting voltage drop across the base-emitter junction and R2 is not adequate to forward bias the two diodes, D1 and D2. This leaves the transistor in a full-on state with the whole supply voltage present at the output terminals.

Now, if a heavy load (0 ohms) is put across the output terminals, the current will incease, but how much? Watch what happens. As the current increases, the voltage drop across R2 also increases. When the base-emitter drop plus the R2 drop reaches approximately 1.2V, the two diodes go into

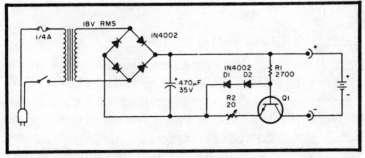

Fig. 8-1. Schematic of universal nicad charger. The transformer is an 18V doorbell type, which results in a rectified 25V to the load.

323

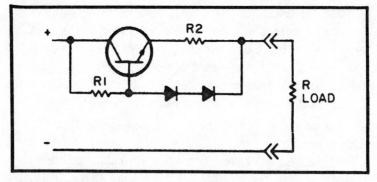

Fig. 8-2. Simplified version of the current limiter in Fig. 8-1. It's just an emitter follower.

conduction and limit any further increase in base potential. Thus the current is limited to that point where emitter circuit voltage drops equal the series turn-on potential of D1 and D2.

For silicon diodes, the turn-on potential is about 0.6V. This also holds true for the base-emitter junction of silicon transistors. This means that the required value for R2 is about 0.6V divided by the current limit desired.

Varying the load (using 1 to 18 nicad cells) reveals that the current limiting action will hold within 1 to 2 mA from 0 to 24V. In other words, you can charge any random number of cells from 1 to 18 without adjusting the charger.

Transistor Q1 should be chosen for a reasonably good current limit value. Since the primary interest is in 450 mA-h penlight cells, the charge current is set at 45 mA. This means that the transistor must dissipate 25V times 0.045A, or 1.125W. Double that for safety and a 2W transistor is about right.

CHARGER WITH AUTOMATIC SHUTOFF

An automatic turn-off feature was designed into the charger so a maximum charge voltage could be obtained. After the predetermined battery voltage level was reached, the charger circuit would automatically turn itself off. Thus, over-voltage protection would be inherent in this design. An over-ride circuit is included in the over-voltage sense circuit. By using an adjustable power autotransformer at the input of the transformer, an adjustable DC voltage is possible at the output when the manual override is enabled.

Figure 8-3 shows the basic block diagram of the power supply with the automatic cutoff feature. The trip voltage for the cutoff is adjustable with a variable resistor. Under normal circumstances, the trip point is preset with a good, fully charged battery connected across the output terminals.

Figure 8-4 shows the schematic diagram of the battery charger. With the output voltage below the trip point level, transistor Q2 is turned on and relay RL1 is closed, applying full output voltage to the charger's terminal. As the battery is charging, its voltage increases.

When the battery voltage reaches the preset level, zener diode D1 conducts, turning on transistor Q2 and the relay. The charger's DC power source is then disconnected from the battery, preventing overcharging. If the battery voltage goes below the threshold voltage by a small amount, the relay will automatically connect across the battery, again charging it. Thus, the charger may be left connected to the battery without fear of overcharging. Resistor R_f is used to provide some hysteresis so the relay does not chatter when the threshold voltage is reached. R_f is chosen for about 0.5V hysteresis. A 12V lamp is used to show when the charger is charging the battery.

To furnish the power source at all times across the charger's terminals, switch S2 grounds the threshold poten-

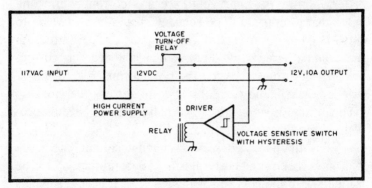

Fig. 8-3. Block diagram of the automatic shutoff charger. Relay interrupts output when the battery voltage reaches the preset threshold.

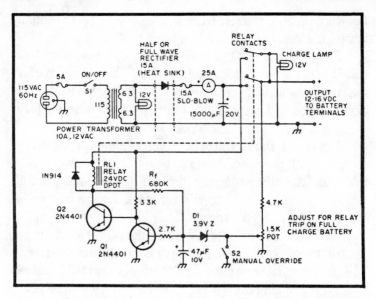

Fig. 8-4. Schematic of charger with automatic shutoff feature. A manual override switch permits the charger to be used by itself for general bench use.

tiometer so the output voltage does not have any effect on the operation of the relay.

If the ripple is too great, a full wave bridge rectifier may be substituted for the 15A diode. The meter may be left out if it is not necessary to monitor the charging current from the charger.

This charging technique can also be used to charge nicad batteries. A resistor should be connected in series with the batteries however, to prevent excessive currents during the initial charge process. The resistor should have a value to correspond to the maximum charging current permitted. As the nicad batteries come up to voltage, the charging current will drop accordingly. When the predetermined battery voltage is reached, the current will go to zero as the circuit is automatically disconnected. The batteries will be thus protected from excess currents.

All of the part values are changed accordingly to agree with one's requirements and junk box. Substitutions should be made to keep the cost to a minimum. The transistors should be NPN types with reasonable current gains. Most general purpose types will work. The 3.9V zener diode can be replaced with a zener diode in the range of 2V to 6V. Its only purpose is to furnish a relatively sharp voltage threshold. The value of the potentiometer may have to be adjusted if the zener voltage is changed drastically. The relay was a 24V DC surplus type capable of 15A of current, but other types should work equally well. Most 24V relays will work in a 12V circuit.

FAILSAFE SUPERCHARGER

A lot of emergency systems employ an automobile battery to power equipment when the 117V AC line fails, but keeping the battery at its peak charge level is quite a problem and is time-consuming.

This is the reason for building the charger. It keeps the battery at full charge and never needs to be disconnected. In fact, since building this charger (Fig. 8-5), a power failure alarm has been added.

This charger is rather simple and doesn't require any expensive parts. Any transformer with a voltage between 14V and 24V and rated at a few amps can be used. The one used here is from an old battery charger.

A two gang rotary switch and two 9Ω 30W resistors are used to give me three charging rates when wired as shown in the schematic. Any charging combinations can be used provided the transformer and bridge rectifier can handle the current. The SCR listed is rated at 7A with a heat sink.

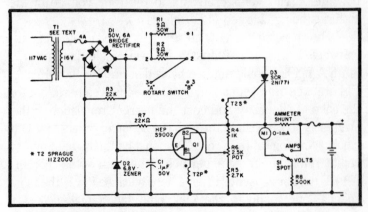

Fig. 8-5. Schematic of failsafe supercharger.

Only one meter is used to read voltage or current. The meter shunt is in series with the positive lead to the battery. Connecting the switch as shown will give you voltage or current. The meter face is reprinted to read 0—15V DC and 0—4A.

The pulse transformer is a small audio transformer which was rewound to give a 1:1 turns ratio at approximately 20Ω resistance. They are sometimes listed as SCR trigger transformers.

The unijunction is used as a relaxation oscillator, and the potentiometer controls the upper voltage limit for the battery. When this limit is reached, the oscillator stops oscillating, which in turn stops triggering the SCR. The battery must be connected for the circuit to operate. If the oscillator circuit fails to operate, reverse one of the pulse transformer windings. The SCR needs a positive spike at the gate to fire.

HIDDEN NICAD CHARGER

A quick glance at Fig. 8-6 reveals the circuit strategy. The details are rather straightforward. The circuit trickle charges the batteries at 40 mA which, in the case of 450 milliamp-hour batteries, is quite acceptable. Some who look at the circuit may balk at the high voltage that appears to be applied to the batteries, but this particular circuit is a variable voltage, constant current charger. Through the loading action, the circuit voltage adjusts to the total voltage of the batteries being charged (12V this case).

The charging current, however, is the variable and critical factor. In the case of constant current trickle charging, 450 mA-h AA cells should not be charged at greater than 45 mA. Elaborate current regulation could be employed here, but for the sake of simplicity and cost the current regulation in this charger is accomplished through the action of capacitor C1. It would seem logical then that C1 should be the highest quality and therefore the most expensive component of the charger. A 1 μF tubular capacitor is used here because it was the most suitable one found in the junk box, but a higher quality capacitor should be used (no electrolytics allowed). The 1 μF

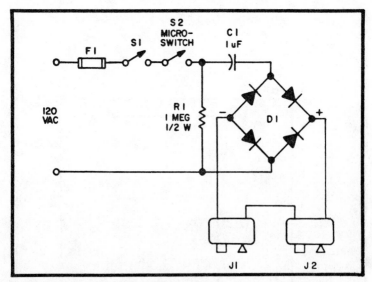

Fig. 8-6. Schematic of hidden nicad charger.

capacitor regulates to about 40 mA of charging current. As a matter of fact, the charging current increases by 4 mA for every 0.1 μF of capacitance used, hence 1 μF equals 40 mA. This variance allows you to choose the right charging current for your need, but remember to stay under your particular charging current limits.

No elaborately filtered DC is needed for charging, so full wave rectifier D1 is selected to do the job. Just about any cheap little full wave package will do, as long as it is rated high enough. A package rated at 200V at better than 0.5A is used here. Resistor R1 is included as a bleeder for C1.

The rest of the circuit is a combination of love for gadgetry and desire for an inconspicuous charger. The entire unit may be installed in a teakwood box that blends into most rooms.

The actual circuit itself (except for switches) can be mounted on a breadboard chip which is placed in a small utility box that can be located at the bottom of the teakwood enclosure. The batteries, along with J1 and J2, are simply allowed to lay on the bottom of the enclosure near the utility box itself.

Precautions were also taken against curious souls opening the box while it was in operation and exposing themselves

329

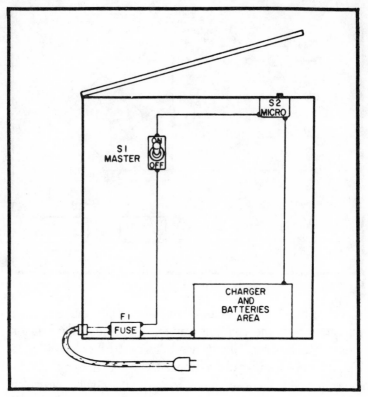

Fig. 8-7. Diagram of charger housing.

to possible shock. This is where switch S2 comes into play. S2 is a microswitch. It is mounted near the top so that closing the lid of the box closes the switch as well. In this configuration the circuit will be deactivated whenever the lid is raised (see Fig. 8-7). S1 is the master switch (optional).

A light bulb or LED indicator could be added. Remember, though, that the circuit voltage will vary from 120V with no batteries attached to 12V while charging. The indicator will have to be designed to compensate, of course. For added safety, all exposed wiring should be well covered.

The other minor details concern F1, J1, and J2. Fuse F1 is a little bit of extra safety to guard against the results of any accidental shorting. The fuse value is not given here since it will vary for most modifications of the design. Each builder should choose a value to suit his own needs.

Jacks J1 and J2 are inexpensive 9V battery clips. They are used as a cheap solution to the accidental reverse polarization problem. It was decided to load the ten AA batteries into one 4-cell plastic AA battery holder and another 6-cell AA plastic holder. The type of holders that sport the mating terminals for the J1 and J2 clips are readily available. Note that they are wired in series. This is necessary for each holder set of batteries to get a full charge.

A few words on the characteristics of the charger itself would be appropriate here. Do not leave the batteries connected to the charger when the unit is not in operation, since the internal resistance of the charger has a tendency to drain the batteries. Also, this charger is a trickle charger, so about 16 hours of charging time works out rather well. As a matter of fact, the batteries can be left charging longer than 16 hours, but such prolonged charging could develop a "memory" in the cells for a particular level of charge which the batteries will not exceed.

AUTOMATIC SHUTOFF NICAD CHARGER NO. 2

Two properties of nicads can be observed whenever they become fully charged. One is the rise in temperature that occurs due to the inability of the cell to make a chemical conversion, which is dissipated as heat. The other is the slight rise in cell voltage from approximately 1.35V to above 1.4V. Whether this rise in voltage is due to the increased temperature or is a physical property of a "flooded" cell quite apart from the temperature is not known. A scheme of placing a thermistor next to the cells to control the charge was abandoned in favor of a means of sensing the voltage increase.

The general circuit is shown in Fig. 8-8. The filtered DC is current-limited by R1 to a value one-tenth of the ampere-hour rating of the battery. Zener diode D1 provides a voltage offset so that all the voltage variation will appear across R2. This pot is adjusted so that the SCR will trigger on the voltage rise that occurs at the end of charge. Whenever the SCR triggers, R3 and R4 shunt most of the current away from the battery so that the battery is getting only a trickle. The LED will light, showing that the end of charge has been reached.

Switch S is used to take the SCR out of conduction and start the charge cycle.

Method

As there are many combinations of transformer voltage and number of nicad cells possible, the calculations are being left up to you; however, the step-by-step design procedure will be demonstrated. Only seven steps are required for your own situation:

1. Specify battery:
   - N = number of cells
   - A = Ampere-hour rating
2. Specify transformer:
   - E = RMS secondary voltage
   - V = 1.414E − 1.4
3. Calculate capacitor value:
   - C = A/120V (farads)
4. Calculate R1:
   - R1 = [V − N(1.4) − 0.7]/(A/10)
5. Calculate offset zener voltage:
   - D1 = N(1.4) − 2.7
6. Calculate R3:
   - R3 = [N(1.4) − 0.7]/[(A/10 − A/100) − 0.01]
7. Calculate R4:
   - R4 = [N(1.4) − 1.8] 100

Example

Assume you have a battery of ten cells rated at 500 mA-h and a transformer rated at 24V (a common voltage available in most junk boxes).

1. N = 10 and A = 0.5.
2. E = 24 and V = 32.5.
3. C1 = 0.5/120(32.5) = 128 μF (use 200 μF, 35V).
4. R1 = (32.5 − 14 − 0.7)/0.05 = 356 (use 360Ω).
5. D1 = 14 − 2.7 = 11.3 (use 11V or 12V zener).
6. R3 = (14 − 0.7)/(0.05 − 0.005 − 0.01) = 380 (use 390Ω).
7. R4 = (14 − 1.8)100 = 1220 (use 1200Ω).

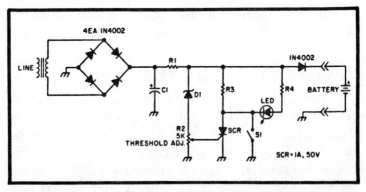

Fig. 8-8. Schematic of nicad charger with automatic shutoff feature.

Construction

If you presently have a desk-top charger for your portable, you can modify the existing circuit. If you're not that lucky, then you may still get a professional-appearing unit by ordering the plastic parts for the charger from the manufacturer. This unit has been built for both a GE-PE and Motorola HT-220. Other portables have a jack for plugging in a charger. In this case, you can build your charger in any of several "project box" enclosures available at hobby electronics stores.

Testing

Substitute a variable resistor for the battery, and, with R2 turned off (wiper at ground), adjust the battery substitute until the desired trip point voltage is reached (1.4V × N). Then adjust R2 until the LED lights and stop there. Now connect your battery and push the switch; the LED should go out. The battery voltage should be between 1.25V and 1.35V per cell during charge and rise above 1.4V per cell at the end of charge.

Parting Shots

You should not trust this charger if ambient temperature is allowed to vary from that which is considered comfortable by most people. Too much heat or too much cold could alter the trip point. You may want to instrument this charger with a milliammeter and measure the various currents for the first few charge cycles, just to gain assurance that it is working okay.

SILVER-ZINC CELL BATTERY CHARGER

The advent of all solid state equipment has resulted in a renaissance for the battery industry. Nickel-cadmium, alkaline, carbon-zinc, and lead-acid batteries are all in widespread use by hobbyists. Due to its cost, the silver-zinc cell has seldom been encountered, although it is highly desirable for really portable operations. This type of cell has been extensively used in torpedoes, missiles, and space applications.

The silver-zinc cell has an extremely high ratio of stored energy to weight, a flat discharge curve and an ability to discharge at a high rate without damage. The cell consists of a silver oxide positive electrode and a zinc negative electrode. The electrolyte is potassium hydroxide. A 40% to 45% concentration should be used. Do not use standard battery electrolyte, which is sulphuric acid. This would immediately destroy the zinc electrode, giving off hydrogen gas.

To place the cell in operation, the electrolyte is introduced through the vent plug. This can be a tricky and hazardous task. For the small size cells normally encountered, the process is best done in the kitchen sink. Remove the small porous filter from under the vent cap. Make a pipette from a piece of glass tubing, or use an old eyedropper that will fit into the cell.

Simply pouring the electrolyte into the vent tube will result in splashing when the cell burps. The cell should be set in the sink or in a glassbowl to prevent damage from the electrolyte. Remember that the electrolyte is highly basic. It can be neutralized with household vinegar.

After filling, let the cell sit for a while, and then top off with more electrolyte to replace that absorbed by the separators. Carefully wipe any spilled electrolyte from the polystyrene case. Charging of the small size cells, with a less than 25A maximum discharge rate, can be readily done with the DC power supply and battery charger whose circuit is shown in Fig. 8-9. If the supply is to be used only for battery charging, the filter capacitor can be omitted. Batteries can readily be assembled from individual cells by placing them in series. Room temperature open circuit voltage for charged cells is typically 1.86V. The normal voltage spread is such that six cells are used for a 9V battery, and 8 or 9 for a 12V battery. Charging current should be limited to between 7% and 10% of the rated cell discharge capacity. This value is generally found as part of the identification number printed on the cell. For example, a Yardney HR-5 series means that the cell is designed for a high rate of discharge, and that the design discharge rate is 5A. In this case, charging current should be kept between 350 and 500 mA. Charging should be stopped when the cell voltage rises sharply and reaches 2.05V (see Fig. 8-10). Do not overcharge, as this will greatly reduce the cycle life.

Manufacturers' specifications showed a dry storage life of up to five years. Packing dates on the surplus cells revealed

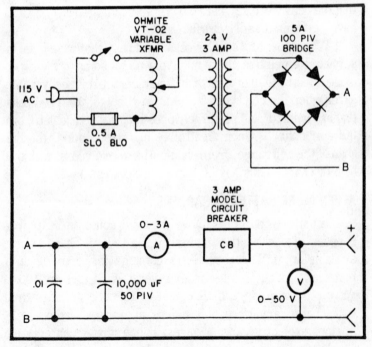

Fig. 8-9. Schematic of silver-zinc cell charger.

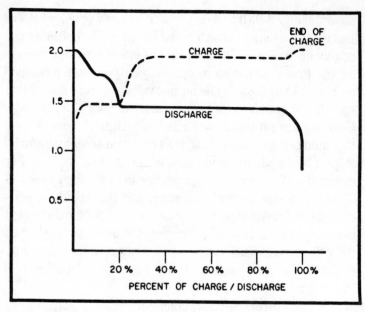

Fig. 8-10. Voltage vs. percent of charge curves for silver-zinc cells.

that the cells were 12 years old, yet they charged right up and have performed satisfactorily.

When selecting surplus cells, avoid any that have been in service. Remember, these might pay off for scrap as the silver oxide electrode has a sizeable percentage of recoverable silver when the cell is discharged. Buy only dry charged cells that are unused, and preferably ones that are still sealed from the atmosphere. Once electrolyte has been added, the life span of the cell ranges from six months to two years, with the high discharge cells having the least life span.

SEVEN MORE BATTERY CHARGERS

A battery charger is one of the most common DC power supplies in use by the general public today. Chargers are available from Wards or Sears catalogs for charging auto batteries at charge rates from several amps to over 100A. Also in common usage are the small battery chargers designed to recharge a variety of cordless appliances, from toothbrushes to electric carving knives. These smaller chargers are usually designed to charge tiny nickel-cadmium (nicad) batteries that are more or less permanently built into the appliances. More recently, the nicad battery has been supplemented by another sealed type of rechargeable battery: the gelled-electrolyte lead-acid battery. Gell-Cell is Globe-Union's trademark for such a battery. The gelled-electrolyte lead-acid batteries are generally of larger physical size and weight than nicads, and are used in portable transceivers, standby lighting systems, instrusion alarm systems, and similar applications.

There are differences among chargers for use with conventional lead-acid batteries (such as used in autos), and still other differences among the small chargers designed for the maintenance-free batteries. We will look at a variety of chargers, to see what they basically do, and to see how they differ.

The main function of *any* battery charger is to cause *current* to flow back into a battery in the opposite direction from which current flowed during discharge. Batteries, when used properly, are relatively constant-voltage devices, so the most meaningful measurements in battery charging are made in *amperes*. The current for charging can come from a number of sources; the most common source, today, is a low-voltage secondary transformer and solid-state rectifier. In years past, motor generators, and low-voltage transformers with older types of rectifiers, were extensively used as chargers. The principle was always the same: force *current* back into the battery to charge it.

The basic battery charger, shown in Fig. 8-11, uses a current source. Since current sources are not as common as voltage sources, an approximation to a current source can be

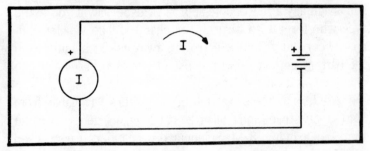

Fig. 8-11. Basic charger using a constant-current generator.

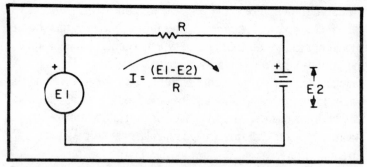

Fig. 8-12. Approximation of constant-current charging using a constant-current generator and series resistance R.

made as shown in Fig. 8-12. By making E_1 *large* compared to E_2 (the battery voltage), essentially constant current will flow during the battery charging cycle, even though E_2 rises slightly as the battery nears full charge. To see quantitatively how this works, let's take the two examples shown in Figs. 8-13A and 8-13B. In both cases, assume that the uncharged (lead-acid) battery starts with an E_2 of 11V, and that E_2 at full charge rises to 13V. In Fig. 8-13A, the current at the start of charge is 5A and at the end of the charge cycle is 4.9A. That is, the charge rate has only changed 2% during the charging cycle.

In the example in Fig. 8-13B, however, current undergoes considerable change during the charge cycle. As in the first example, the charge starts out at 5A, but drops to 2.5A at the end of the cycle. This latter case is typical of most inexpensive battery chargers on the market; the number of ampere-hours put into the battery is difficult to calculate, however. This latter case is a simple form of tapered charging and is actually a reasonable approach, especially if you're the type who forgets to turn off the charger. The smaller end-charge current will not electrolyze away as much water from the battery as fast as with constant-current charging (if the charger is left on too long).

Modern commercial battery chargers, available from auto supply stores and mail order catalogs, are generally of the form shown in Fig. 8-14. Selenium disc rectifiers are still used because they have a rather large forward voltage drop and so

338

can make it unnecessary to use a dropping resistor. Another trick used in commercial battery chargers is to wind the transformer so as to have a relatively large equivalent secondary leakage inductance, thus eliminating the need for a current-limiting resistor.

In older types of chargers, copper-oxide-disc and tungar-bulb rectifiers were used; these rectifiers also had considerable forward drop. Older chargers can occasionally be retrieved from the "back of the garage" and returned to use by replacing nonfunctioning rectifiers with modern silicon rectifiers; however, when doing this replacement, be sure to add some current-limiting resistance to make up for the higher efficiency of the new diodes. The value of resistance can readily be determined by charging a discharged battery, starting with a reasonable guess as to resistance value, and watching the ammeter. One can even add the current-limiting resistance in the primary of the charger transformer, if that's easier. Primary current-limit resistors are, of course, higher in resistance than those in the secondary circuit. A homemade

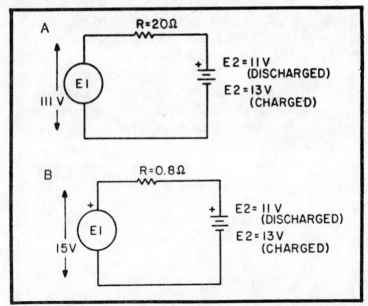

Fig. 8-13. Examples of constant-current charging. (A) With E_2 discharged, $I = (111 - 11)/20 = 5A$. With E_2 charged, $I = (111 - 13)/20 = 4.9A$. (B) With E_2 discharged, $I = (15 - 11)/0.8 = 5A$. With E_2 charged, $I = (15 - 13)/0.8 = 2.5A$.

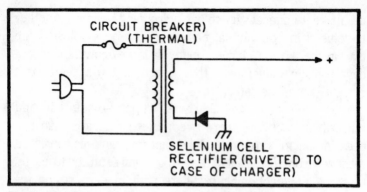

Fig. 8-14. Typical inexpensive commercial battery charger.

charger using a 200W light bulb as such a primary resistor is shown in Fig. 8-15.

In recent years, there has been progress in making electronically controlled chargers that charge conventional lead-acid batteries completely automatically. Such chargers are now commercially available, and you may connect your battery to the charger and essentially forget it. Such an automatic charger is the Heathkit GP-21. The GP-21 is not only self-controlling, but is protected against shorting the battery leads together and reversing the battery leads. Either of these two operator errors produces no sparks or opening of breakers, but simply nonoperation. The Heathkit GP-21 is shown in Fig. 8-16. Note that Q3 and Q4 are involved in the lead-shorting and reversal protection; the basic charger can be simplified (for circuit understanding) to that of Fig. 8-17. Whether the GP-21 charges the battery depends on whether the 12.8V

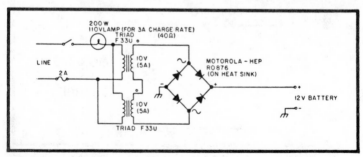

Fig. 8-15. Schematic of simple battery charger with current-limiting resistor in the transformer primary.

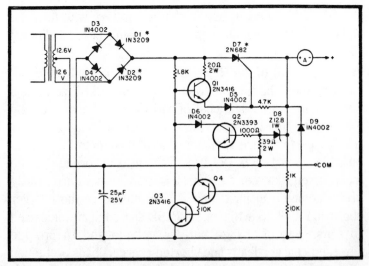

Fig. 8-16. Heath GP-21. Components marked by asterisk are all on one heat sink.

zener (D8) condusts; that is, whether the battery terminal voltage is over 12.8V. If the battery voltage is below 12.8V, the battery need charges, D8 does *not* conduct, Q2 is off, and Q1 is allowed to pass full-wave rectified 60 Hz via R1 and D5 to the gate of D7. Thus D7 turns on (at 120 Hz rate) to pass current to the battery. If, on the other hand, the battery voltage is above 12.8V, D8 conducts, causing Q2 to be on, which clamps the base of Q1 to ground (via D6) and full-wave rectified 60 Hz is *not* passed via R1, Q1, and D5 to the gate of

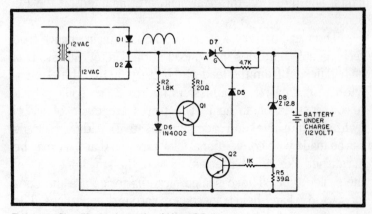

Fig. 8-17. Simplified schematic of Heat GP-21.

341

D7. So in this second case (a charged battery with terminal voltage over 12.8V), D7 is prevented from passing additional current to the battery.

The circuit of the Heath GP-21 could probably be duplicated by many home constructors, but the 12.8V zener is a nonstandard value. The cost of parts would almost certainly exceed the Heathkit cost, however. An automatic charger that has performance similar to that of the Heath GP-21 was built that uses two 6.3V, 10A, filament transformers as the main current source; such filament transformers are usually pretty common in junk boxes. The feature of the charger shown in Fig. 8-18 is that an LM319D dual comparator is used to sense end-of-charge battery voltage and to provide reversed-charger-leads and shorted-charger-leads protection sensing. In this circuit, the end-of-charge voltage is *adjustable* by means of a small 2K pot. The LM319D uses a ±15V supply which is provided by a Triad F40X, bridge rectifier, and Raytheon RC4195 dual regulator IC. A fraction of the +15V regulated voltage (as voltage divided by the two 6.2K resistors in series) is used as the voltage reference against which the lower comparator compares a fixed fraction of the battery voltage. The function of the upper comparator is to sense a reversed-leads or shorted-leads condition. In order to facilitate construction of this charger, the control circuitry has been laid out on a PC board, shown in Fig. 8-19. The larger components, and those on the heat sink, are mounted off the PC board.

The nicad battery was one of the first maintenance-free rechargeable batteries introduced in the United States. It is quite different from the lead-acid battery, in that the electrolyte is a strong base (potassium hydroxide or sodium hydroxide) instead of an acid. The plates are made of nickel hydroxide (positive) and cadmium (negative). Although nicads can be made with conventional filler caps, so that they may be maintained like auto batteries, most of the small ones are of the sealed type. These sealed nicad batteries are the ones most used in hand-held transceivers and small appliances. The electrolyte in such batteries is held in a separator between the

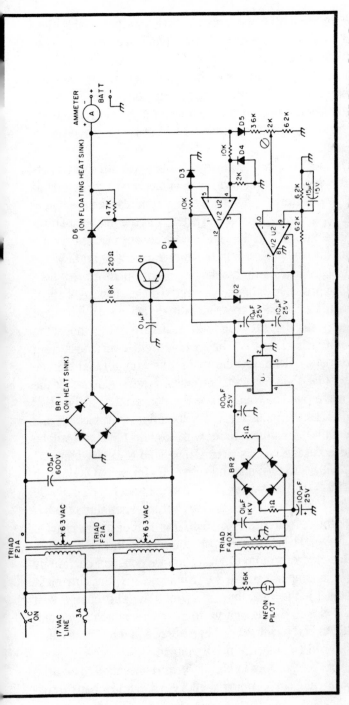

Fig. 8-18. Schematic of battery charger using IC control with adjustable finish voltage. Diodes D1 through D5 are 1N4002 or HEP R0051. Diode D6 is a 2N682 or HEP R1471. Bridge rectifier BR1 is a 12A integrated bridge Motorola MDA980-2 or HEP R0876. Bridge rectifier BR2 is a 1A integrated bridge Varo VE27. Transistor Q1 is a 2N3641 or HEP S0015. Chip U1 is a Raytheon RC4195NB. Chip U2 is a National LM319D.

343

positive and negative plates, which is also gas permeable. The gas permeable separator absorbs oxygen which the battery liberates when it is overcharged. Almost all sealed nicad cells have a safety release valve, however, so that in extreme overcharge conditions they won't explode. Some of these valves are of the resealing type and some are not. The point here is that even sealed nicads may vent gasses, if severely mistreated.

The sealed nicad is generally charged with a constant-current type of charger, with the addition of a voltage limit across the battery. This sort of charger can be relatively simple, as several construction articles in the hobbyist literature have shown. In Fig. 8-20, three approaches to the constant-current, limited-finish-voltage charging method are shown. The first method relies on the nonlinearity of light bulbs to approximate constant current (the resistance of a light bulb decreases with decreasing voltage across it). The light bulb charger is shown in Fig. 8-20A. Figure 8-20B shows the FET approach. The FET has an inherent constant-current characteristic which is utilized in this charger. The large number of FETs is required because it is desired to utilize the inexpensive plastic types. Since the nominal I_{dss} of the HEP F0015 FETs is nominally 2 to 20 mA, some picking and choosing must be done. Units with I_{dss} of 8 to 15 mA will be considered usable in this circuit. The sum of I_{dss} for units Q1 and Q2 will be 15 mA, and the sum of I_{dss} for units Q3 through Q6 will be 35 mA. A switch is used to connect Q3 through Q6 for a total current of 50 mA during charge; when this switch is opened, the current drops to 15 mA for float or trickle charge. Zener diode D1 limits the finish voltage, as in Fig. 8-20A.

The charger in Fig. 8-20C produces constant current by means of a simple transistor circuit, in which the current is adjustable by the user by means of R1. The charge-float switch is closed during charge to raise the value of constant current. As in the chargers of Figs. 8-20A and 8-20B, a zener diode is used to limit the finish voltage.

All of the chargers in Fig. 8-20 were intended to be used with a ten-cell pack of nominal 12.5V, as used in many trans-

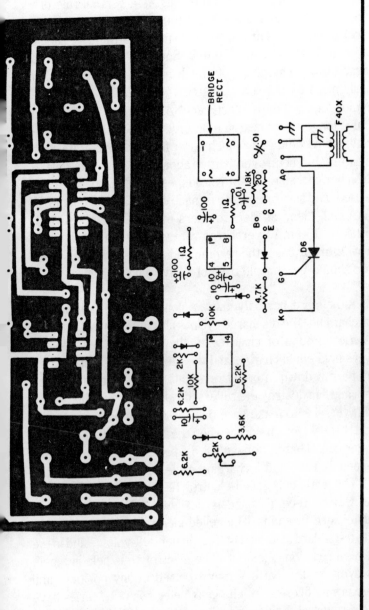

Fig. 8-19. PC board for the charger in Fig. 8-18. (A) Foil layout. (B) Parts placement.

345

ceivers. The individual cells in such a pack are of the AA-size (penlite) and have about 400 mA-h (milliamp-hour) capacity. The Sonotone S-101 or General Electric XGCF450ST are representative units of this type cell. As a general rule one charges at a rate that is one-tenth the milliamp-hour rating of the cell or battery. That is, for a 400 mA-h battery, we should charge at a rate of about 40 mA. Such a charge rate would appear to fully charge a 400 mA-h battery in 10 hours, but usually nicad batteries are overcharged at least 25%, as recommended by the manufacturer. No damage is supposed to occur if longer over-charging occurs, but most battery manufacturers recommend a much lower float or trickle rate after the battery has been overcharged about 150%. It is possible to use higher charge rates with nicads, especially if special types are used, and accurate timing is made certian. For example, in an AA-cell, General Electric offers the XKCF450ST, which will take a 150 mA overcharge rate (more than three times normal). In any case, the recommended rates of the manufacturer should be consulted before any rapid charge mode is tried.

Although it has only indirect connection with nicad battery charging, the discharge cycle is also quite important, too. In series groups of nicad cells, the individual cells do not always hold equal charge, and the one which discharges first may be run down into cell reversal. This is considered bad form, and can destroy the reversed cell. One way to help avoid cell reversal is to make sure that the system powered by the nicad battery stops its drain when the battery voltage drops to 1V per cell. That is, for our 12.5V pack, operation would be stopped at 10V, and a charging cycle started.

The lead-acid, gelled-electrolyte battery has been introduced more recently, and is designed to be a lower cost maintenance-free unit. The gelled-electrolyte battery is similar to a standard lead-acid battery in per-cell-voltage and in the chemical reactions inside it. The electrolyte is held in a gell, however, so the battery may be used in any position, and there are no filler caps. The gelled electrolyte batteries have no cell reversal problem nor the memory effect characteristic

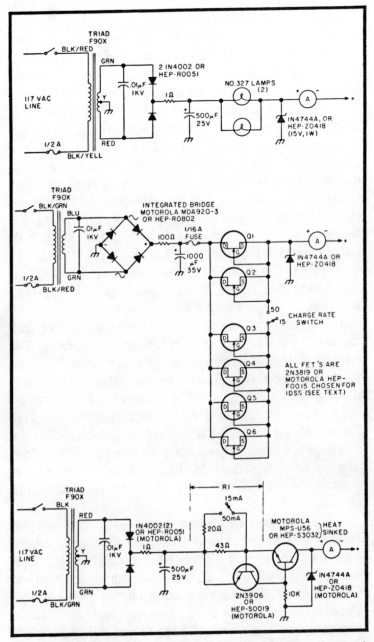

Fig. 8-20. Three typical chargers. (A) Light-bulb type charger. The resistance of the light bulb changes as the voltage across it varies. (B) FET-type constant-current charger. The FET has an inherent constant-current characteristic. (C) Transistor constant-current charger.

347

of nicad batteries. They do vent during normal overcharge, and usually have resealing vent valves for this reason. This venting causes a loss of the water in the gelled electrolyte, which cannot be replaced, so the cells do have a limited life, depending on how much overcharge is used. The recommended method of charging one such gelled-electrolyte battery, the Elpower Solid Gell battery, will serve as an example of how at least one manufacturer feels it should be done. The Elpower EP1230A is a 12V, 3 A-h unit and is charged at a maximum current of 0.45A until a battery voltage of 14V is reached. The voltage is then held constant at 14V until the charge current drops to 0.040A. At this point the charger is disconnected or switched to float. The float mode keeps 2.2V per cell across the battery, or 13.2V for the EP1230A.

The charger in Fig. 8-21 is designed to charge a 12V, 3 A-h gelled-electrolyte battery, such as the Elpower EP1230A. It is essentially a constant-voltage regulator with current-limiting as designed around a National LM305H, with the usual PNP-NPN pair to increase current capability. The constant output voltage is adjusted by means of the 5K pot and the current limit is controlled by the value of the 0.68Ω resistor. This 0.68Ω resistor is actually a 1Ω and a 2.2Ω half-watt resistor in parallel. All the circuitry above the dotted line is added to the standard regulator to accomplish the special gelled-electrolyte charging requirement. This added circuitry consists of 1N4454, LED, LM311H voltage comparator, 2N4302, FCD810 optocoupler, six resistors, and two pots.

The way that the charger circuit works is as follows. When you connect a discharged battery across the output, the battery will tend to draw unlimited current, since the supply has constant voltage output, and the battery voltage is (presumably) lower. The current-limiting function of the LM305H immediately comes into effect, however, and only 0.45A is passed, so the output voltage drops to that of the battery. As the battery charges (at the constant-current rate of 0.45A), its terminal voltage rises slowly to 14V, but cannot rise above this voltage because of the *voltage* regulating action of LM305H circuit. So when the battery voltage gets up to 14V, the charging current starts to diminish. During the initial constant-current and constant-voltage phases of charge, the voltage at point A is higher than that at point B, both with reference to point C. Voltage BC is constant because of the regulator formed by the 2N4302 and 1N4454. As charging current through the 0.68Ω resistor drops, voltage AC drops

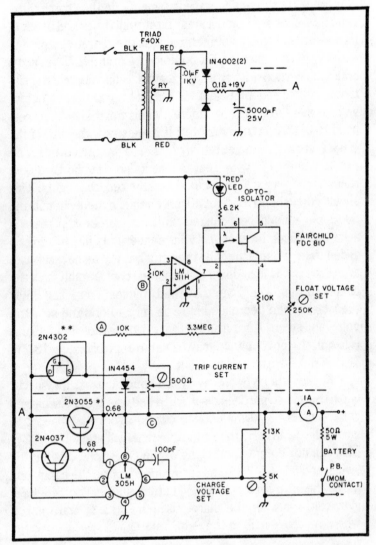

Fig. 8-21. A 12V constant-voltage charger for gelled-electrolytic batteries.

until it is no longer more positive than voltage BC. At this point the LM311H output changes from high to low, which turns on the LED *and* the LED inside the optocoupler, FCD810. This in turn causes the optotransistor in the FCD810 to saturate, and connects the 10K and 250K pot across the LM305H sensor divider. The constant-voltage output of the LM305H regulator is thereby dropped from +14V to +13.2V.

Since there are several adjustments on the charger, the setup is somewhat complicated. First, pull the FCD810 out of its socket, and with a voltmeter across the charger output (no battery, yet), adjust the 5K pot for +14V output. Then add a 350Ω, 25W rheostat across the terminals and decrease resistance until regulation is lost (as indicated by a drop in output voltage as read on the voltmeter). This point should correspond to 0.45A on the ammeter; if it does not, the size of the 0.68Ω current-sense resistor should be adjusted. The FCD810 should now be restored to its socket and the 350Ω rheostat increased in resistance until the red LED lights. This should correspond to about 40 mA on the ammeter; if not, the 500Ω pot should be adjusted until the proper trip point is achieved. Since the LM311H has deliberately had hysteresis added to its circuit (the 3.3M resistor), it will be necessary to adjust the 350Ω load down in resistance considerably until the circuit untrips, as indicated by the LED going out. The 350Ω rheostat is then decreased in resistance to find the new trip point, and so on. After a satisfactory trip point of about 40 mA is found, the float voltage may be set to approximately 13.2V. This is done with the 250K pot.

Finally, there is one awkward condition that may occur when charging a partially discharged battery. If the discharged battery does not draw enough current from the charger to untrip the LM311H, the output voltage will stay at +13.2V. This will not do much charging of the battery. A pushbutton switch and 50Ω, 5W load are provided to untrip the circuit and move the charger to a +14V output state. The red LED indicator is on when the charger is in the +13.2V state and off when the charger is in the +14V state.

Chapter 9
Sensing Circuits

This chapter contains some rather unique circuits for the power supply builder. These sensing circuits include types that meter both current and voltage. Since it's current that kills and causes the smoke, we are chiefly concerned with the amperes.

VOLTAGE LIMIT SENSOR

The voltage limit sensor (VLS) is a compact, self-contained go/no-go indicator which tells at a glance if the voltage in an automobile or boat electrical system is satisfactory. Many of the latest state-of-the-art electronics equipment have incorporated into them various sensors which continuously monitor test points. Whenever one of these test points deviates outside prescribed limits, a warning light or indicator of some type alerts the operator. In the go/no-go variety of indicators, similar to the oil pressure, generator, and temperature lights on automobiles, only a critical condition will necessitate some action on the part of the human operator. The low cost sensor described here provides an alerting indicator if the voltage in the electrical system falls outside safe limits.

Theory of Operation

Operation of the sensor is very straight-forward. Referring to the schematic diagram of Fig. 9-1, the voltage input provides both the sense voltage and the supply voltage needed to operate the VLS.

The undervoltage part of the VLS consists of Q1, Q2, DS1, and D1 along with three resistors R1, R2, and R3. When the input voltage is less than the breakdown voltage of D1, transistor Q1 is turned off. This in turn allows the current flowing through Q1's collector load resistor, R3, to flow into the base of Q2. This causes Q2 to go into saturation acting like a switch to light the amber indicator, DS1. As the input voltage goes through the zener breakdown point, current begins flowing into the base of Q1. When Q1 has gone into saturation, no base current is available for Q2, which turns off. Indicator DS1 also goes out since Q2 is cut off. Resistor R1 assures a sharp turn on of Q1 and also provides a path for collector leakage of Q1. R2 limits the base current into Q1 after D1 is conducting.

The overvoltage part of the VLS consists of Q3, DS2, D2, and resistors R4, R5, and R6. Q3 is cut off until the input voltage exceeds the zener breakdown of D2. At this point, base current flows into Q3 turning on indicator DS2. R4 and R5 serve similar functions as R1 and R2 described above. R6 is a current limiting resistor so DS2 does not burn out for the higher input voltages.

Design Criteria and Construction

Silicon transistors are used to assure stable operation over a wide range of temperatures. Transistors, Q1, Q2, and Q3 were chosen to have high beta of 100 to 300, and a high collector current rating of at least 100 mA. Voltage breakdown can be 20V to 25V or more. A collector power dissipation rating of 300 mW or greater is also desirable.

The voltage at which an automobile operates its primary low voltage system is a function of temperature. For example, a typical GM product will have a range from 13.5V at 150°F to 15.2V at 0°F. The combination of zener diodes D1 and D2 plus the small base to emitter voltage drops of Q1 and Q3 were chosen such that any voltage less than 13.5V would light the amber indicator and any voltage more than 15.2V would light the red indicator. The 5% zener diodes assure an accurate turn on and turn off without any adjustments.

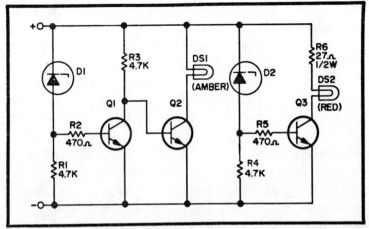

Fig. 9-1. Schematic of voltage limit sensor.

Since the VLS detects a voltage falling outside these defined limits, it was felt that tracking as a function of battery temperature was not justified. An elaborate temperature sensing circuit was deemed unnecessary and beyond the basic requirements of the VLS.

The circuit is constructed in a small Bud minibox without crowding. The two pilot lamp indicators are mounted in one end and a rubber grommet in the other end for the two wires. If the VLS is to be used on a negative ground system and the unit is to be securely fastened to the metal ground of the automobile or boat, then the negative lead can be grounded to the case internally and only one wire, the positive lead, brought out of the unit. No special wiring precautions are necessary; however, if sockets are not used for the transistor it is recommended that a heat sink be used on the leads during the soldering operation. This will prevent the possibility of damage to the transistors from excessive heat.

Checkout and Installation

Since the VLS draws a negligible amount of current during normal operation and only 80 mA during the time an indicator is on, power can be obtained from almost any point in the low voltage electrical system. However, it should be switched on and off with normal ignition and accessories since with just 12V input, the amber indicator will be on drawing continuous current.

During operation at an ambient temperature of about 75°F where the voltage input will be about 14.2V or so, it is possible to use a 1.5V dry cell battery placed in series with the positive lead to check the VLS. With the 1.5V battery positive terminal connected to the VLS (battery voltage adding), the red indicator should light. With the 1.5V battery negative terminal connected to the VLS (Battery voltage subtracting), the amber indicator should light. This test will generally work unless the automobile low voltage is not adjusted properly or the ambient temperature is very high or very low. In these cases, a bench-type variable-voltage power supply could be used for final checkout.

The voltage limit sensor will monitor your 12V battery and charging system alerting you only to potential unsafe conditions not indicated on the usual idiot light.

EASY CURRENT TESTER

Here's a really easy way to measure current flow. Since most transistorized gear is powered from a battery pack of some sort, all you have to do is stick a piece of double-sided printed circuit board between any two batteries, as shown in Fig. 9-2.

Touch the two meter leads to each side of the PC board and there you have it: instant current reading without cutting and soldering any wires.

TWO POWER SUPPLY LOAD TESTERS

The first question asked might be, "What is a load bank?" The term is taken from the electrical power industry to explain a device to simulate a load on various power sources in order to check the load performance of those sources. In this case, it covers the building of two simple load banks or dummy loads for small and rather large sized DC power supplies that are finding present usage with hobbyists. This section shows how it can be done economically and with readily available parts. Also, using the basic ideas presented here, load banks for most any type of low-voltage power supply can be built.

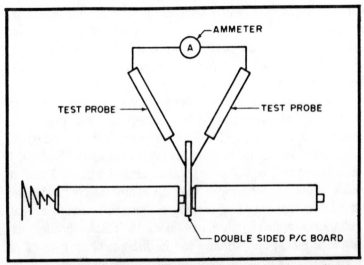

Fig. 9-2. Method of measuring current in battery operated equipment. Insert a double sided PC board between two cells. Touch the test probes of the ammeter to each side of the board and read the current flow.

In order to properly evaluate the performance of a power supply (whether home built or commercially built), the supply must be tested at various degrees of loading—that is, no load, half load, full load, etc. The major requirement for a suitable load bank is that it be capable of providing these various load conditions. Thus, several load elements are required, or at least one large variable element is required, in order to cover several different load conditions. These load elements would typically be large fixed or variable power resistors. If the power supply is of any size, the load elements have to be of high power rating (10W, 25W, or even 50W). A look at a parts catalog will show that power resistors get expensive as the power rating goes up and the different ohmic values available go down. Also, power resistors present the problem of how to get rid of the heat and how to conveniently mount the resistors.

Instead of using the regular power resistor in the design of these load banks, automotive bulbs and pilot lamps were used. These are readily available at auto parts stores, electronic parts houses, and even department store electrical parts sections. Bulbs are selected on the basis of voltage and current requirements and typically work out very nicely in regard to both values. The problem of power dissipation is very minimal and mounting can be very simple. With the load elements decided, the actual construction of the load bank can begin.

The first load bank requirement was for a rather large 13.8V 24A DC supply. It was decided to check one-third, two-thirds, full load, and 133% of full load. Voltage and ripple were observed under the various load conditions. The #1073 automotive bulb was selected with a nominal rating of 12.8V at 1.8A. Typical cost is less than 30 cents. Since the power supply voltage was to be a volt higher, the current at that voltage was approximated to be about 2A. This is a fine feature of these bulbs; the voltage can be increased up to 20% of nominal rating without any problem in this application.

The bulbs were arranged in four groups or sections of four each. See Fig. 9-3 for the circuit diagram. Each group presents a load of about 8A at nominal voltage to simulate four load conditions. In order to keep costs down and construction simple, the wires consisting of AWG #14 TW were soldered directly to the bulb bases. The start and end of each section wire were wrapped around a nail which was driven into the wooden base board. This eliminated expensive sockets and, after all, a load bank is not a device that is normally on display.

Since this load bank was to be used for a considerable amount of testing and since four surplus relays were available, each load section was wired through relay contacts as shown in Fig. 9-3. Any combination of loading can be quickly selected. This extra feature can be eliminated with just the use of solder connections or switches with proper current rating. The switching feature is certainly handy if considerable testing is to be done, but adds considerably to the parts cost if the junk box is not well stocked.

The second load bank was for a much smaller DC power supply with a rating of 5V 1A. Following the same design features of the previous load bank, a #502 bulb was selected as load element in four sections with two bulbs in each section, giving 30%, 60%, 90%, and 120% of full load. Since the

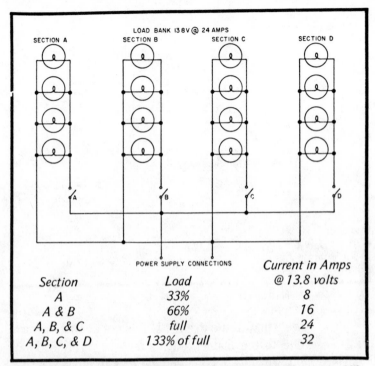

Section	Load	Current in Amps @ 13.8 volts
A	33%	8
A & B	66%	16
A, B, & C	full	24
A, B, C, & D	133% of full	32

Fig. 9-3. Schematic of power supply load tester rated for 13.8V and up to 32A.

current requirements were much less, about 0.3A per section, regular AGW #22 solid hookup wire was directly soldered to the bulb bases and toggle switches (or even slide switches) used for controlling the load sections. A small wood base board was again used for mounting the bulbs and wiring with a small ⅛-inch pressed board panel (nailed to the edge) to support the switches. This supply is most useful for checking logic power supplies used with ICs. Overload or current foldover characteristics can be checked on supplies designed with that feature. See Fig. 9-4 for the circuits.

There are many advantages to these simple load banks besides the cost and ease of construction. The bulbs present a large amount of light under full load to leave no doubt that the supply is working. Also, there are no burn marks from power resistors on the workbench or, even worse, the dining room table. Buying the bulbs by the box reduces cost, and automotive type bulbs are readily available, even at gas stations.

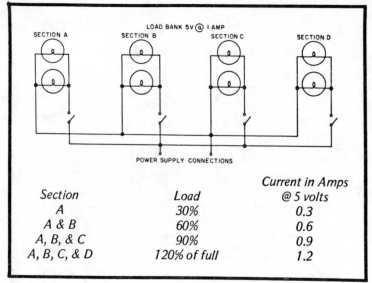

Fig. 9-4. Schematic of power supply tester rated at 5V and up to 1.2A.

Section	Load	Current in Amps @ 5 volts
A	30%	0.3
A & B	60%	0.6
A, B, & C	90%	0.9
A, B, C, & D	120% of full	1.2

The purists may argue that the bulb is not a constant load resistance due to the filament characteristics. This sudden heavier than normal load when voltage is first applied to the bulb is very short in duration and provides an even stricter load test of the supply being tested.

The load bank ideas presented here can be extended to other sizes and types of load banks by changing the bulb types and the number of bulbs in a section, and even the number of load sections.

These load banks have been used to check voltage and ripple conditions on a variety of home brew and commercial supplies. The cost is certainly cheap if proper shopping around is done on the choice and source of bulbs.

THE SMOKE TESTER

This tester allows full power (smoke) testing of large or small power supplies *before* they are put on the rig. It will tell how many volts at how many amps the supply can deliver, and what kind of voltage loss to expect under load. It can also help determine some of the attributes of low-voltage power transformers, without having to wire them into a power supply.

It consists of two or three transistors, a pot, a couple of meters, a few diodes and some loose hardware. Don't forget the heat sink. If you wish to be formal about it, then add a chassis of some kind.

Figure 9-5 shows the basic circuit. As the wiper is moved toward the plus, more current flows in the base circuit. The external circuit looking into the collector-emitter junction sees a lower and lower resistance. The ammeter and voltmeter give an idea of watts cooking (sorry about that). The diode allows less than perfect attention to the polarity of the incoming voltage. The resistor in series with the base limits base current to a safe value.

Although the transistor manual indicates that the 2N3055 is a 15A transistor, two or three of them cleared like fuses when the collector current was slowly increased above about 11A or 12A. The 10A fuse was put in the emitter, and collector current excursions were limited after that.

Since power supplies capable of giving more than 20A were to be tested, two 2N3055s were put in parallel. (See Fig. 9-6.) The fuses gave enough resistance in the emitters to help the transistors share the current in a more or less equal manner (at least there were no more creamed transistors).

A Darlington or piggyback configuration is used to reduce the power dissipation requirements of the pot. Ten amps in the collector can require as much as 0.5A in the base. Half an amp times several volts equals several watts. The Darlington system reduces the input current by as much as 50 times. That

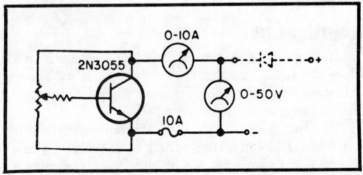

Fig. 9-5. Schematic of basic smoke tester. This verison is only capable of 10A.

359

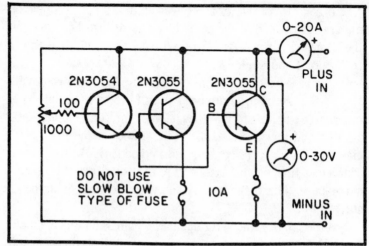

Fig. 9-6. Schematic of beefed-up smoke tester capable of handling 20A.

figures out to about 10 to 20 mA for 10A to 20A in the collectors, and makes the system practical.

Note that if the power supply is delivering 15A at 13V, then something is going to have to dissipate 195W. *Be sure* that the 2N3055s are on a large enough heat sink.

If a full-wave bridge rectifier (Fig. 9-7) is placed in front of the tester, power transformers can be checked for voltage out versus current. That information would have saved a lot of time in the past.

Of course, there are single transistors around that meet the voltage, current, and power ratings of this tester.

If PNP transistors are used, reverse the polarity of the meters and diodes.

AMMETER SHUNTS

This is a simplified method for cutting ammeter shunts. There are two things that will not happen. It will not be necessary to do any real deep math, and the meter will not get pinned during the cutting process.

The voltage drop along a piece of wire is proportional to its length and the current through it. If a piece of heavy wire is hooked up as shown in Fig. 9-8, all of the sport is taken out of making meter shunts.

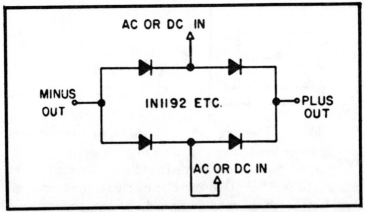

Fig. 9-7. Bridge rectifier. If a bridge is placed in front of the smoke tester, you can check out power transformers for voltage versus current.

A power supply, ammeter, resistor, and the bus bar that is about to become a shunt are all connected in series. Long thin flexible leads from the 0–5 milliammeter can be used. After all, they are going to carry no more than 5 mA. The negative lead is tied to the power supply and bus minus.

First the power is turned on and adjusted to give the required current. This is shown on the standard VOM. The plus lead of the 5 mA meter is touched to the bus *near* the minus lead. There should be a small deflection seen on the meter. The lead is then slid along the wire until full scale deflection is obtained. That point on the wire is marked.

If a single range is all that is required, it only remains to solder the leads in place. The resistance change at the point of soldering will be small in comparison to the resistance of the meter. Therefore, the soldering operation will not upset the calibration.

Leave a little more bus at each end than is needed. If the shunt is rather long, then slide insulation over it before the

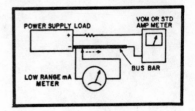

Fig. 9-8. Setup for shunting ammeter.

361

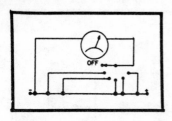

Fig. 9-9. A multirange ammeter.

final soldering. The wire can be wound on a convenient coil form in order to make it somewhat more compact.

If a 0 to 20A range is needed and there aren't more than 2A available, or the VOM doesn't have a high current range, all is not lost. Set things up as shown and adjust the power supply for, say, 2A. Then slide the plus lead for one-tenth (twenty divided by two) of full scale. This will generally prove to be adequate calibration if the standard is reasonably close. Later, when the high current supply is finished and a higher range standard is available, it will take almost no effort to touch up the calibration.

If the shunt turns out to be 10 inches long for 500 mA, then halfway there, or 5 inches up from the cold end, will give a 1000 mA, or 1A, shunt.

A multi range ammeter (Fig. 9-9) was constructed with this method in less time than it used to take to make a single shunt. Again, since the meter and its leads are carrying only 50 μA or 5 mA (whatever the basic movement is), then a cheap multiposition switch and light hookup wire will do the job. AWG #22 wire may be used for shunts up to about 1A.

Chapter 10
Automotive Accessories

Today you'll find that some people have as many electrical gadgets in their cars as they do in their home. All types of power are required and many problems are encountered. Here's some handy information about alternators, regulators, and the like. If the answer you're looking for is not here there's those other three chapters on AC-to-DC converters and DC-to-AC inverters, and DC-to-DC converters.

THE ALTERNATOR

While there are a number of alternator/regulator systems, some characteristics are common to all of the recent models. To begin with, all of the alternators are designed to be mounted on the engine block; they are belt driven. The alternator pulley is about 3 inches in diameter while the driving (crankshaft) pulley is not less than 6 inches in diameter. As a result, the alternator shaft will turn at a speed that is at least twice the speed of the engine; a commonly found ratio is 2.5:1. Rotation is normally clockwise as viewed from the front (pulley end) of the alternator. It is important that the alternator be rotated in the proper direction; reverse rotation will cause the integral fan to move less air and the alternator may overheat if it is run at full load.

Alternators use a rotating field to which DC is supplied through a pair of slip rings and carbon brushes. This arrangement permits the high output currents to be supplied directly from the stator windings without going through brushes or sliding contacts. Field current is usually less than 3A for alternators that are rated at less than 50A output. The rotating field is built with six pair of poles and so the output of any one stator winding goes through six electrical cycles with each revolution of the alternator shaft. The output frequency in hertz is equal to one tenth of the shaft speed expressed in revolutions per minute. For example, if the alternator shaft is turning at 4000 RPM, the output frequency will be 400 Hz. Exception: Delco-Remy alternators used in GM cars generally have seven pairs of poles and produce seven cycles per revolution.

The stator has three windings and it supplies 3-phase power; the windings may be connected in either the delta or wye configuration, the wye connection being the most common. The stator leads are connected directly to the internal rectifier which is made up of six silicon diodes. The diodes are mounted in the alternator and are arranged to provide full-wave rectification. The ripple frequency is six times the frequency developed in any one winding. At a shaft speed of 4000 RPM, the ripple on the DC output will have a frequency of 2400 Hz (2800 Hz in GM cars). The rectifier diodes may be mounted on the rear end bell of the alternator or on two separate plates (or a printed circuit board) mounted inside of the rear end bell. Three of the diodes are built with their cathodes connected to their cases; the structure that they are mounted on is connected to the positive output (BAT) terminal. The other three diodes have their anodes connected to their cases which in turn are grounded to the alternator frame. Several makes provide a capacitor that is connected between the output terminal and the frame to protect the diodes from voltage surges; it also acts to suppress radio noise. A typical alternator schematic is shown in Fig. 10-1.

The DC output is a function of the shaft speed and the field current; an increase in either one will raise the output voltage. The alternator is self-limiting, however, in that when the shaft speed exceeds about 5000 RPM, the output does not

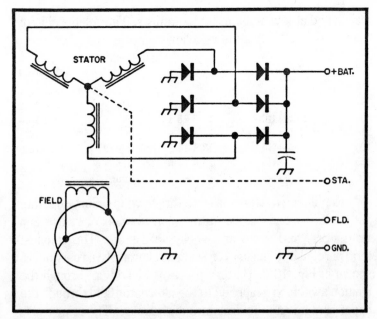

Fig. 10-1. Typical alternator schematic showing Y-connected stator windings. The stator midpoint is brought out to a terminal only on Ford units.

continue to rise in proportion to increased speed. This effect is mainly due to the fact that the flux created by currents induced in the stator windings opposes the flux created by the rotating field. Designers call the phenomenon armature reaction. Hysterisis losses in the stator also contribute to self-limitation. In most current models, the alternator output is regulated by changing the current in the winding of the rotating field.

The alternator output terminal is directly connected to the battery in all cases. Direct connection is possible because the reverse leakage through the diodes in the rectifier (less than 1 mA) is so small as to be insignificant. This arrangement eliminates the necessity to provide a relay having heavy contacts capable of disconnecting the alternator from the battery when the engine is not running as is done with DC generators.

Regulating Systems

The regulator is made sensitive to the voltage of the auto's electrical system. To a lesser extent it also reacts to the ambient temperature, with slightly higher voltages being

365

maintained at low ambient temperatures. The relationship of temperature to voltage is as follows:

Temp. °F	Chrysler	Ford	GM	Motorola
0	–	–		14.6-15.4
25	–	–	Range	14.4-15.2
50	13.6-14.6	14.3-15.1	is 13.5	14.2-15.0
75	13.5-14.5	14-1-14.9	to 15.2	14.0-14.8
100	13.4-14.4	13.9-14.7	for all	13.8-14.6
125	13.3-14.3	13.8-14.6	temper-	13.6-14.4
150	13.2-14.2	–	atures.	13.4-14.2

Regulating systems fall into three groups. Some operate with relays alone, some with a combination of relays and transistors, and some are wholly solid state. The simplest form of regulator consists of a single relay and two resistors as shown in Fig. 10-2. Battery power is picked up through the ignition switch. It is applied to the alternator field through the upper contact and the armature of the relay. The relay coil is connected between the ignition system terminal and ground. The action of the relay can be compared with that of a voltmeter, the armature acting like the indicating pointer of the meter. The voltage required to just pull the armature away from the upper contact would be the nominal voltage listed in the listings above. The additional voltage necessary to pull the armature to the lower contact would be from 0.2V to 0.6V greater than the voltage required to pull the armature away from the upper contact. When the electrical system voltage is low, the relay armature rests against the upper contact and the full system voltage is applied to the field winding.

As the battery becomes charged, the system voltage increases and the relay armature is pulled into a position between the upper and lower contacts. The opening of the upper contact places a resistance of about 8Ω in series with the field, causing the field current to drop from some 2.5A to about 1A. A further increase in the electrical system voltage will cause the relay armature to close against the lower contact. In this condition the field lead is grounded and field current will drop to zero. A resistance of about 20Ω is provided to absorb surges generated in the field winding and thus

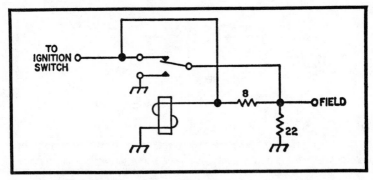

Fig. 10-2. Schematic of a single-relay regulator.

protect the relay contacts. In normal operation the armature
will first rest against the upper contact for a short time after
the engine is started. At moderate speeds it will vibrate
against the upper contact and at road speeds it will vibrate
against the lower contact (assuming the battery condition to
be normal). The vibrating relay switches the alternator condition rapidly between full output and partial output or between
partial output and no output. The rate at which the vibrations
occur and the length of time that the relay armature rests
against either contact is determined by the system voltage and
the alternator response characteristic for the speed at which it
is turning. While the average system voltage should be within
the limits given in the previous listings, the instantaneous DC
voltage (other than ripple) measurable in the electrical system
will vary between 11.5V and 15V. It will change with each
vibration of the relay armature, which might be as high as 100
Hz or so. Single-relay regulators provide no means of operating a charge indicator or idiot light; an ammeter must be used
to show whether the battery is charging or discharging.

Some GM and Ford automobiles employ a two-relay
regulator system. The two relays function in a way that permits the use of either an ammeter or a charge indicator lamp. A
typical two-relay regulator is shown in Fig. 10-3. Some early
Chevrolets employ a third relay, separately mounted, to operate the charge indicator lamp. In later models the relay was
made an integral part of a three-relay regulator. Some GM
autos are equipped with a relay-type regulator that contains a

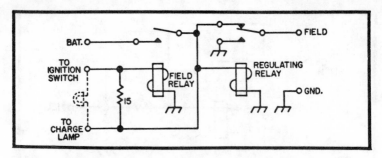

Fig. 10-3. Schematic of a two-relay regulator with a terminal for operating a charge indicator lamp.

power transistor which isolates the field coil from the relay. This arrangement contributes to longer relay contact life, making a more reliable regulator. One of Ford's arrangements uses a connection to the wye-connected stator neutral to operate a separately mounted field relay; the balance of the regulator is transistorized and contains a variable potentiometer for voltage adjustment. A significant characteristic of relay control is that there is a level of generator output between zero and full output, with the charging rate controlled by switching the generator rapidly between two of the three conditions.

Several auto manufacturers have been using solid-state electronic regulators. Included are American Motors, Checker, Ford, GM, and several truck makers. The solid-state regulator offers a more stable and reliable control of the alternator output. It is free of the maintenance problems that arise from the aging of relays, pitting or wear of contacts and the effects of dirt accumulation.

A schematic of a typical transistorized regulator is shown in Fig. 10-4. The regulator consists of two PNP transistors, a zener diode, and an assortment of resistors. The regulator circuit is a two-state, high-speed switching network. The values of the components are chosen so that a voltage below the selected operating potential, the zener diode is nonconducting. The resulting positive potential that is applied through R3 to the base of driver transistor Q1 biases it to cutoff. With Q1 cut off, only a minute amount of current flows through R6 and the base of Q2 will be nearly at ground potential, biasing it to full conduction. In this condition the positive potential of the electrical system is applied to the alternator field through R7.

As the electrical system voltage increases, there is an increase in the potential that appears between the base of Q1 and the junction of R2 and R3. A zener diode having a reverse breakdown or zener voltage of from 8V to 10V (depending on the manufacturer) is connected between these two points. When the zener breakdown voltage is reached, the diode conducts and the positive potential on the base of Q1 is reduced. Q1 then conducts, raising the voltage developed across R6 and biasing Q2 to cutoff. This interrupts the flow of current in the alternator field, causing the alternator to stop developing power. As the system voltage drops, the zener diode stops conducting, Q1 is again biased to cutoff, Q2 conducts, and current flows to the alternator field. The switching action takes place so rapidly that it can go through as many as 2000 switching cycles per second. The key to transistorized regulator operation is the zener diode which acts both as a voltage reference and as a voltage actuated switch which initiates regulator action.

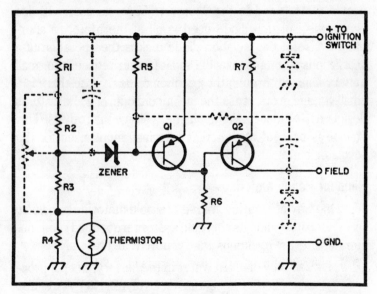

Fig. 10-4. Schematic of a typical solid-state regulator. Components shown connected by dashed lines are used by some manufacturers but not all of them.

It is possible to add temperature correction to the regulator by using a thermistor. A thermistor is a special kind of resistor with a very pronounced negative temperature characteristic. Its cold resistance can be several times its hot resistance. The zener diode senses system voltage through the network of resistor R1 through R4. The thermistor is connected across R4. As the temperature rises, the resistance across R4 becomes less and a greater proportion of system voltage appears across the zener diode. The net result is that the system voltage will be regulated at a slightly lower voltage when the ambient temperature is raised.

A few components are shown in the schematic that should be discussed. Ford and GM regulators are equipped with a potentiometer in place of resistors R2 and R3 to permit a small range of regulator voltage adjustment. They also use a feedback circuit from the collector of Q2 to the base of Q1 to speed up the switching action, together with diodes, intended to protect the regulator from surges appearing on the field or ignition leads. GM also places a capacitor across the combination of R5 and the zener diode to smooth out the voltage surges in the diode circuit. Figure 10-5 shows the delta connected stator used in the Motorola 55A alternator. It also shows the isolation diode used in all of the Motorola alternators. Use of the isolation diode makes the operation of a charge indicator lamp possible and it further reduces the small battery leakage through the rectifier diodes. Since there is a small voltage drop across the isolation diode, the alternator is designed to produce a slightly higher voltage than others. This voltage is present at the AUX terminal and is used for the regulator and field supply.

Hints for Happy Motoring

One way of assuring yourself reliable motoring is to *know* that your battery and its charging system are in good conditon. Here are few suggestions that you may find helpful:

1. Use only distilled water in the battery; keep it properly filled. Check the electrolyte level at regular intervals and after periods of heavy charging.

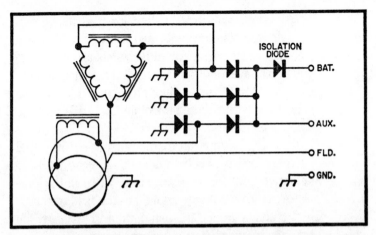

Fig. 10-5. Schematic of a Motorola alternator showing the delta-connected stator used in 55A models. Also shown is the isolation diode and auxiliary terminal connection.

2. Measure the specific gravity of the electrolyte occasionally; when fully charged it will measure 1.265 ± 0.01 at 80°F. Correct for temperature by adding 0.004 to the measured value for each 10° that the electrolyte temperature is above 80°; subtract the same amount for each 10° that the electrolyte temperature is below 80°.

3. Provide a means of giving the battery a supplemental charge when it is needed; such occasions include periods after extensive winter night driving, when electrical usage has been heavy, and whenever the generating system is not in top form.

4. Regulators sense system voltage; if possible, equip your car with a good voltmeter. Surplus aircraft style voltmeters and ammeters are still available at reasonable prices.

5. If you have equipped your car with a voltmeter, an ammeter, or both, consult them often enough to know which conditions are normal and which are abnormal.

6. Make this simple test occasionally to verify that your charging system is in good shape: with the engine running at the speed it would attain at the road speed

of 30 MPH, turn on the headlights (high beam), electric windshield wipers, and heater fan. That will place a load on the electrical system of about 25A. (If you don't have electric wipers, use the cigar lighter; it draws about 10A.) Under these conditions the charging system should maintain 13.5 to 14V at the battery terminals.

7. If your alternator system will not put out its rated current, look for such troubles as a loose fan belt, defective contacts or blown fuse wires in the regulator; test the voltage applied to the field terminal. If the above items are okay, then investigate the brushes for wear. The alternator bearing lubrication and diodes should be checked last since testing the diodes requires that the alternator be disassembled and the diode leads be unsoldered.

Some Tips for Experimenters

Automotive alternators can be arranged to supply 6V or 12V DC, 60 Hz or 400 Hz power to emergency rigs. A gasoline engine rated at 2 or 3 horsepower makes an ideal prime mover. Keep your eyes open for discarded power lawn mowers if you want a cheap engine. Used alternators are easy to find at most auto wrecker's yards for a few dollars. Both can be mounted on a piece of 1-inch lumber, leaving room for the regulator or field supply arrangement. The engine speed and alternator shaft speed can easily be optimized by using pulleys of the proper sizes and a V-belt. The alternator will be most efficient when operating at speeds in the range of 4500 to 5500 RPM. Small gasoline engines will develop their rated power at some speed between 2000 and 3000 RPM; this speed is often listed on the name plate or in the instruction manual.

Even the smallest alternator made should be able to develop at least 350W of electrical power. Most alternators are Y- connected and will furnish two DC voltages simultaneously if the common connection or neutral is brought out. (Ford Autolite alternators bring this point out to a terminal designated STA.) The voltage across the arms of a

Y-connected three-phase generator or transformer is 1.73 times the voltage developed in one winding. The DC at the stator common is produced by three-phase, *half-wave* rectification, however, and so its ripple frequency will be only half that present on the regulator DC output (BAT) terminal. Losses in the diodes, added to the above conditions, result in a DC voltage at the stator neutral that is approximately half that at the BAT terminal.

Residual magnetism in the alternator field is often so low that the output voltage will not build up until the field is excited by an external 12V source. Once the alternator is running and developing power it can be kept self-exciting. If the alternator is to supply DC to the radio equipment, it is a good idea to leave the storage battery connected to smooth the ripple in the alternator output. With such an arrangement, any type of automotive alternator regulator should work satisfactorily. If the alternator is used to supply AC, it is better to use a rheostat for excitation control to avoid the on-off operation of the regulator. One such arrangement is shown in Fig. 10-6. The two rheostats are 10Ω 50W units and are adjusted to produce the proper field current while limiting the battery charging current to about 1A. The lightest possible DC load is recommended to avoid excessive flattening of the output waveform. The alternator output is a fairly good approximation of a sine wave unless it is supplying a heavy DC load. A heavy DC load will cause the output waveform to be severely clipped. When the alternator is self-excited it is in a positive feedback situation, i.e., as the field current goes up, the

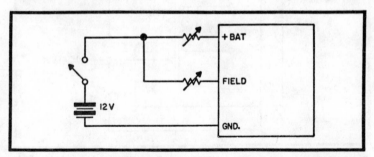

Fig. 10-6. Arrangement for adjusting the field current to a constant value when the alternator is used to supply AC.

output voltage goes up, which in turn results in more field current. Under no-load conditions the alternator voltage could get high enough to damage the rectifier diodes, so when experimenting, bring the field current up from a low value rather than down from a high value.

Alternators can be used to supply 60 Hz equipment. When turning slowly enough to develop 60 Hz power, the alternator will be pretty inefficient and will require more field current than is normal. It is best to push the speed (and frequency) as high as the radio equipment will allow. Most radio equipment of good design will operate well on frequencies around 100 Hz if the power is supplied in the manner shown in Fig. 10-7. This kind of connection can be easily made to any receiver that is equipped with a socket for the supply of external power. Notice that filament power is supplied directly

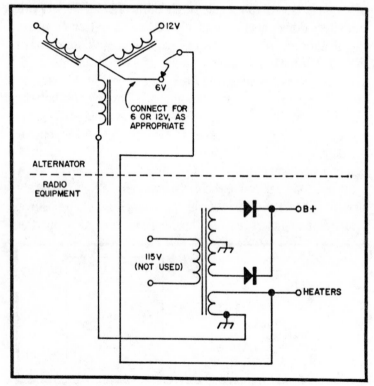

Fig. 10-7. A simple way to operate radio equipment from low-voltage alternator AC.

from the alternator; the filament winding becoming the transformer primary. In this condition the transformer is called on to supply only that amount of power that is required for supply voltage. The reduction in total power transferred by the power transformer is one of the reasons that the equipment tolerates the higher frequency so well.

The arrangement shown in Fig. 10-7 can also be used well with 400 Hz equipment. This is a particularly promising field for experiment since 400 Hz power transformers of the highest (military spec.) quality are available at rock-bottom prices. They are a lot lighter than 60 Hz transformers, too. When using the arrangement in Fig. 10-7, be sure to isolate the alternator frame electrically from the radio equipment. The more ambitious experimenter may want to isolate one alternator winding for use with the rectifier for field supply and use the other two for AC supply. While much of the foregoing presumes that the alternator will be driven by a small engine, it can also be mounted on an engine block as a special supply for mobile operation. Watch for wide voltage swings if this is done, and provide for protection of 400 Hz equipment against excessive currents that can flow if the frequency drops radically.

INSTALLING AN AMMETER

In an effort to simplify the operation of automobiles, most car manufacturers have, for many years, incorporated an alternator indicator light on the instrument panel in lieu of the ammeter of days gone by. While this indicator light requires less attention while you are driving, it does not give you complete information on the state of your electrical system.

Advantages

The system described here has several advantages which make it attractive. These advantages are:

1. All leads are at ground potential, so there are no problems with shorts in the system and there is no fusing the conductors.
2. No heavy currents pass through the indicator, so small conductors can be used.

3. Although the system makes use of shunt resistance to measure the current, a unique method is used so that no additional voltage drops are added into the present electrical system.
4. It is very easy to add to your existing electrical system with no changes in existing wiring.
5. The ammeter circuit can easily be adapted with a switch and resistor to read system voltage.
6. It is inexpensive to install and, depending on the type of meter movement you install, the cost can range anywhere from $1.50 to $7.00.

With these advantages, it is felt that more people will want to install an ammeter to their car electrcial system to monitor the battery charging rate.

Typical Electrical System

Referring to Fig. 10-8, the Ford electrical system incorporates an electromechanical voltage regulator and is the type used in many cars; other systems such as General Motors are similar. As you can see, the alternator consists of a stationary three-phase stator winding in which the output goes to a set of diode rectifiers. These diodes are arranged in such a manner that the output at the battery terminal of the alternator is always positive. There is a rotating field which excites the stator winding during normal operation. As you can see, the alternator is always connected directly to the battery through a terminal block, and it is at this point that the other electrical loads in the car are connected. The neutral from the stator winding is grounded through the field relay winding in the voltage regulator; its contacts are normally open.

When the ignition switch is turned on, a certain amount of voltage from the battery is connected directly to the field through the charge indicator light on the instrument panel. After the engine is started and the alternator is producing current, the stator current flowing through the field relay winding pulls the contacts down, which in turn bypasses the indicator light on the instrument panel. In other words, as long as the alternatoɩ is producing current of sufficient magnitude

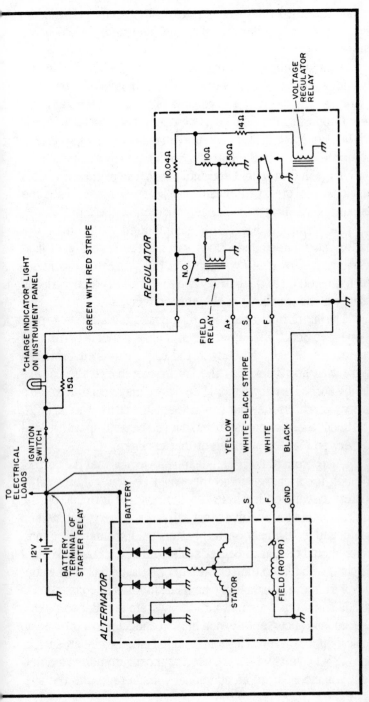

Fig. 10-8. Ford electrical system with electromechanical type regulator. Other electrical systems are similar.

377

to hold the field relay contacts closed, the charge indicator light will be off, indicating that the alternator is producing current.

This system works fine as long as all elements in the system are operating under normal circumstances; however, as long as the alternator is capable of producing enough current to operate the field relay contacts, the charge indicator will be extinguished although there may not be enough current produced to charge the battery. Under these circumstances, the charge indicator light does not give a true picture of what is happening in the system, and it also does not indicate the amount of discharge or charge condition of the battery.

Referring to Fig. 10-9 a simplified diagram of the battery-alternator connections without the regulator of field windings is shown. As you can see, the alternator is continuously connected to the battery and to the other electrical loads usually at the terminal of the starter relay.

During normal operation of the car, the alternator performs two functions. First, it produces enough current to operate the electrical loads in the car. Second, it produces current to flow back into the battery to charge the battery during normal driving conditions. There may be a time when a faulty alternator does not produce enough current to do both. It is under these circumstances that you can end up with a dead battery and no forewarning of that condition.

Also, from the diagram, it appears logical that the correct location for a charge indicator would be somewhere in the battery circuit, either between the positive terminal on the alternator and the positive terminal on the battery, or between the negative terminal on the battery and ground. This indicator should be a zero center scale so that it would indicate a discharge condition (battery drain) or indicate a charging condition from the alternator. The usual method of connection is to add some type of shunt into the system in order to sample the amount of current flowing in the conductors to provide an indication; however, if a shunt is added into the circuit, such as the 0.25Ω shunts which are used on most automotive ammeters, you incorporate an additional voltage drop into the sys-

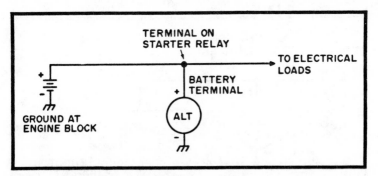

Fig. 10-9. Simplified diagram excluding the regulator.

tem. By accident, a unique feature was discovered of automobile systems which you can make use of what is called a built-in shunt condition.

Ground Paths

Referring in Fig. 10-10, a simplified electrical diagram indicating electrical loads connected to the system is shown. This is a discharge condition in that the engine is stopped and there is no output from the alternator. Since the alternator remains connected to the system, it would appear that some drain would also result from the alternator; however, keep in mind that the alternator windings are connected to the positive terminal through diodes. Therefore, no reverse current can flow through the alternator. Under this condition, the alternator current $I_A = 0$. Applying Kirchhoff's law which states the sum of the currents entering a junction is equal to the sum of the currents leaving the junction, there is the battery current I_B entering the terminal and the load current I_L leaving the junction. Since the alternator current is equal to zero, $I_B = I_L$, or simply stated, the battery is supplying all of the required power to operate the load.

Let's discuss for a minute the theoretical versus actual conditions. We normally think of the negative terminals of the battery, alternator, and loads all being rounded together so that all negative terminals are at the same potential; however, in most cases, the frame of the car is ground return path. Since steel is a mediocre conductor, there is always some inherent

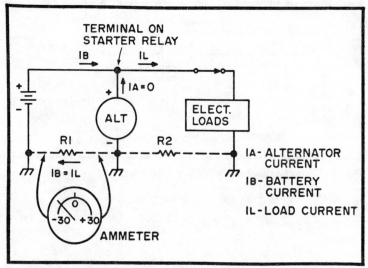

Fig. 10-10. Basic electrical system showing resistance in ground path circuit in discharge condition. Engine is stopped and electrical loads are connected.

resistance incorporated into all automobile electrical systems. This resistance is represented as being R1 between the battery negative terminal and the alternator negative terminal and R2 between the alternator and the electrical loads. Applying Ohm's law, the voltage drop across R1 would be:

$$E = (I_B) (R1)$$

Granted, the resistance of R1 is small, in the neighborhood of 0.01Ω to 0.1Ω, but the currents are in the neighborhood of 10A to 30A (except during starts when it is much higher). Consequently, the voltage drop across R1 is a measurable quantity and can be used to indicate which way the current is flowing in the battery-alternator loop. This is shown on the meter in Fig. 10-10 as a discharge situation.

Figure 10-11 is a diagram of a basic charge condition in which the engine is running and no loads are connected. Again, applying Kirchhoff's law with $I_L = 0$ we have: I_B is equal to I_A, and the voltage drop across R1 is $E = (I_L) (R1)$, but is of the opposite polarity of the discharge condition in Fig. 10-10. As the battery becomes charged, the alternator current charging the battery gradually tapers off and so will the voltage drop across R1. Therefore we get a direct indication of the condition of the battery, and we know exactly when it has reached full charge condition.

In Fig. 10-12 we have a diagram showing normal operation of the vechicle in which the engine is running and certain electrical loads are connected. Under this situation, again applying Kirchhoff's law, we have I_B and I_L leaving the terminal and the alternator current. I_A, entering the terminal. Stated as an equation, $I_A = I_B + I_L$. This means that the alternator has to supply enough current to operate the electrical loads and charge the battery at the same time. Of course, as the battery becomes fully charged again, battery current will be operating mainly to supply current to the electrical loads connected to the system.

Ohm's Law Applied

It is apparent that there will be some voltage drop which can be measured across this ground path resistance, but just exactly how much will be measured and how can it be utilized in a charge indicator?

Let's say, for example, that the ground path has a resistance of only 0.01Ω and that a typical alternator is capable of producing at least 30A. The voltage drop across the ground

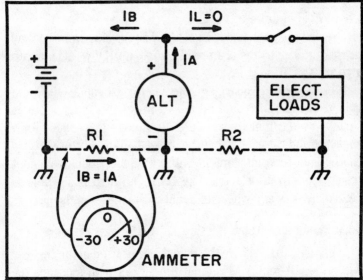

Fig. 10-11. Charging condition with engine running and electrical loads off.

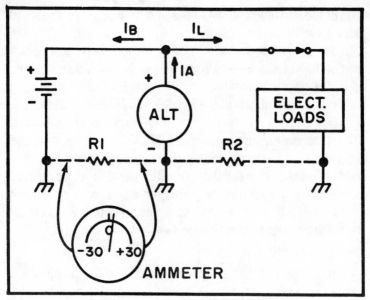

Fig. 10-12. Typical operating condition with battery charging current tapering off.

path of resistance would then be:

$$E = (30)\ (0.01)$$
$$= 0.3$$

where E is in units of volts.

This voltage is easily measured and the instrument may be calibrated in terms of amperes, that is, 0.1V would indicate a current of 10A.

If the normal charging rate is, for example, 10A, the voltage measured would be 0.1V. From this example, it is clearly evident that there is plenty of voltage available for a sensitive indicator that can be used to our advantage in a charge indicating instrument without adding additional shunt resistance into the electrical system. Our next problem is to determine how and where to connect the meter in the circuit.

Basic Ammeter Circuit

Figure 10-13 shows the actual circuit used in the test car with a 50 μA meter. This instrument, as with most instruments (even the very inexpensive tuning indicator instruments) can be converted to a zero center scale instrument so that the meter can become a ±25 μA meter. A meter protector circuit should be installed across the terminals to limit the current during starting. Also adding a 2K resistor into the circuit allows for calibration of the meter.

As mentioned, the basic movement that should be used is a 0 – 50 μA meter manufactured by Midland, Model No. 23-206; however, any meter may be used, possibly up to a 0 – 1 mA meter, depending on your own particular situation. You may want to use a VOM to check your particular ground path circuit under normal conditions to find out how sensitive a meter movement will be required in your particular case. After this has been determined and you have a meter available, the next step is to carefully remove the plastic cover on the meter. Then, after removing the two retaining screws of the dial, carefully slip the dial out from under the pointer. Then, all that is necessary is to erase the lettering not required and to add in new lettering according to your own particular circumstances.

The next step is to convert the instrument to a zero center scale type instrument. This is easily accomplished by moving the zero adjustment until the pointer is at midscale. It is possible to do this on all instruments which have been checked so far, including the little horizontal movements that are available through surplus outlets for $1.50. The quality of the instrument is governed only by your own particular needs. The final step is to replace the dial and plastic cover and install

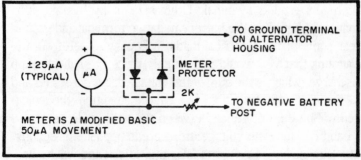

Fig. 10-13. Actual diagram for adding an ammeter to car electrical system. No changes are required to existing wiring.

the protective diodes (Fig. 10-13) across the meter terminals along with the multiplier resistor used for calibration.

Installation Details

The next step involved installation in the car and connecting the meter leads to the proper points in the electrical system. It is best to use solid conductor wire of approximately AWG #18. If you can find Teflon-coated conductors, so much the better. Since the leads will be near hot parts of the engine, Teflon-insulated conductors are more resistance to heat and grease in this type environment.

The two leads from the instrument circuit should be routed through any available opening in the fire wall and around to the alternator and negative terminal of the battery. One conductor is connected to the ground terminal on the alternator housing. The other conductor is connected to the negative battery terminal right at the right polarity; leave the engine off and turn on some load such as headlights or press on the brake pedal. The instrument should swing from the zero center scale to the left indicating a discharge condition. If it swings the other way, the leads should be reversed.

The next step is to get a rough calibration on the instrument. Turn on headlights or some other load with a known ampere rating to roughly calibrate the instrument based on this amount of load. If you have another ammeter, you can use it to calibrate your new instrument.

It is imperative to connect the ammeter circuit into ground path R1 between the battery negative terminal and the alternator ground terminal. Otherwise, if by mistake it is connected between the battery negative terminal and ground near some of the electrical loads, it is easy to see from the diagrams that you would get roughly twice the indication on discharge, which is the sum of the voltage drops across R1 and R2 in a discharge condition. In a charge condition, the indication will be the difference between the voltage drops across R1 and R2 since the voltage drops are in opposite directions. This obviously will give you false indications and will be useless as far as determining the state of charge of your battery.

Conclusion and Results

You are probably wondering at this point as to whether or not it would have been easier to buy one of the commercially available ammeters for $6 to $10 and not have the problem of converting another meter for use in this project. It is true it may have been easier, but the meter movement in commercial automotive instruments is not known for its quality, and there are usually no means of calibrating the commercial instrument. Also, most commercially available instruments make use of an additional meter shunt into an electrical system of approximately 0.25Ω. Admittedly, this is not very much resistance to add into a circuit, but 0.25Ω at 20A could introduce an additional voltage drop into the automobile electrical system.

The test charge indicator has been in operation for some time, and it has been a great help in monitoring the automobile electrical system.

ANOTHER AMMETER FOR THE CAR

Most people belabor the fact that many of the present day cars come equipped with those questionably functional idiot lights. The average person not qualifying as an idiot, finds them extremely difficult to read in determining how much current his accessories are drawing from an already overloaded battery.

The generally accepted method for monitoring current in a vehicle is to mount an ammeter somewhere in or on the dashboard, break the DC line at an appropriate place, run a pair of heavy leads through the firewall and connect them to the meter.

This has many disadvantages. One time trying this using AWG #10 insulated copper wire, a voltage drop of approximately 0.7V under a load of 18 to 20A was noted, not to mention the job of knocking a large size feedthrough hole in the firewall. It is suggested that you take a lazy-type approach to this problem of remotely reading current.

This is accomplished by the simple expedient of inserting a zero center 30 − 0 − 30 microammeter from the negative battery terminal to any point on the body of the vehicle where

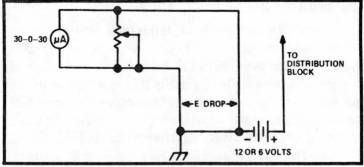

Fig. 10-14. A simple ammeter for your car.

a good ground can be obtained (Fig. 10-14). This has the effect of reading the voltage drop across the negative battery cable which is directly proportional to the current flow.

In the test case it was found that the meter reading was slightly higher than the actual current. This was proved by inserting a standard 0 − 30A meter in the DC line and observing the current when the light or other accessories were turned on. An ancient wire wound, open type potentiometer was scrounged from the junkbox and placed across the microammeter terminals. This was adjusted until both meters read the same. The ammeter was then removed and the DC line closed.

A 10-foot piece of AWG #18 insulated wire was then connected to the negative battery terminal. This was guided through the cable hangers and through a convenient small hole in the firewall where it was connected to one terminal of the microammeter, the other terminal going to ground. The preset pot was soldered across the terminals. The meter was then mounted in a small panel designed to attach to the steering post.

I now have a device with which I can read current charge and discharge with a fair degree of accuracy.

Remember that this scheme involves wiring that is virtually at the same potential as the vehicle body, therefore the danger of short circuits is non-existent. The main advantages of this simple installation are: elimination of voltage loss at the load; and ease of installation.

SOLID-STATE ALTERNATOR REGULATOR

The regulator described here was designed mainly for use with a 40A or 100A Leece-Neville alternator, but the circuit is compatible with just about any alternator made. If you are using the 100A unit, omit the current limiting pot and connect your bias wiring directly to the output of the regulator. The 100A jobs have a tremendously low source resistance and current limiting protection is never required; however, fuses and advisable.) If you have a mechanical regulator, the best place for it is the round file. These gadgets offer poor regulation and always choose to fail when you're in Timbuktu. Moreover, it would cost twice as much for a mechanical unit which would do the same job as this transistorized unit.

In the test car the regulator performs as follows: the system voltage from the alternator does not vary more than 0.1V from 0 to 70A. From 70 to 120A the voltage drops a total of 0.25V. At 150A out, the voltage is down only 0.34V from the nominal setting. Running the alternator over 100A for a period exceeding 5 seconds is not recommended. The alter-

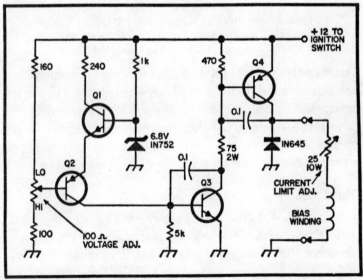

Fig. 10-15. Circuit diagram of the solid-state automotive voltage regulator. This circuit will provide excellent regulation to over 100A. Although it was designed to use with Leece-Neville alternators, it will work with just about any make. Transistor types are listed in text.

nator will take it, but the rectifiers are not only expensive, they are cumbersome to replace.

Construction

The regulator described here (Fig. 10-15) was built on a sheet of fiberglass breadboard material; however, construction is not critical, so build it in or on anything you like. Make sure to heat sink Q3 and Q4. Be sure the heatsinks are exposed to free air. Heat sink Q3 as though it were to radiate 2 or 3W and Q4 10 to 20W, and you will be in good shape.

The circuit is designed so that you can use either germanium or silicon transistors anywhere in the circuit. The listing below shows recommended transistors of both types.

Transistor	Type	Silicon	Germanium
Q1	NPN	2N1711 1	None recommended*
Q2	PNP	2N1132	None recommended*
Q3	NPN	2N657	None recommended*
		2N3738	
		2N3739	
Q4	PNP	2N3790	2N174, 2N1537,
		2N3789	2N3611, 2N3612

*No germanium transistors are suggested for Q1, Q2 and Q3 because excellent silicon units should cost less than $6.00. If you use silicon all the way through, the cost should be less than $13.00 total.

Transistors Q3 and Q4 will no doubt be the most expensive, so use germanium if the cost of silicon is prohibitive. Under no circumstances use germanium transistors, however, if you drive in an extremely hot climate or through deserts. Silicon transistors can withstand two to four times the temperature of a comparable germanium type.

Installation

Once constructed, install the regulator under the hood, perferably in front of the radiator so that it is cooled by virgin air rather than warmed air. On a hot day the temperature under the hood of a car is typically 200°F. Connect the output of the regulator (ground and the collector of Q4) to the bias winding of your alternator. This winding is sometimes referred to as the excitation coil.

Time to connect 12V. Select a point that is turned on and off by the ignition switch, fused and easily capable of delivering 3A. Such a point is obtainable at the main fuse block in your car. Be sure to provide a good ground for the regulator.

Adjustment

Connect an accurate voltmeter across the battery. Throttle the engine up to about 1200 RPM and adjust the regulator for a system voltage of about 13.8V. If you are using the current limiter pot, be sure it is in the shorted position before making the above adjustment. There is nothing magic about the number 13.8; in fact, you should use the lowest value that will still maintain a good charged condition on your battery. After establishing the above potential, it is time to adjust the current limiter pot if you are using it. Turn on everything you normally use in the car: lights, radio, etc., but do not exceed the maximum current you feel your particular alternator can safely deliver. Now, with the current limiter pot, adjust for a system potential of about 12.4V. Further current will now be drawn from the battery instead of the alternator.

Circuit Analysis

All high quality voltage regulators depend on a difference amplifier of some sort. In this case Q1 and Q2 serve the save purpose but are complementary; Q3 and Q4 further amplify and increase the open-loop gain of the regulator to an extremely high value. The alternator in the feedback path feeds back a value of EDC to the voltage divider in the regulator. Figure 10-16 demonstrates how this system resembles a DC feedback amplifier. An ordinary feedback amplifier is also shown.

If the ratio of R1 to R2 is quite small compared to the open-loop gain of the amplifier (amplifier gain without feedback), then the gain will, in fact, nearly approach the ratio R1/R2. At this time one can appreciate the illustration in Fig. 10-16. It should be evident that if the open-loop gain is tremendously high, the regulator (amplifier if you like) will not

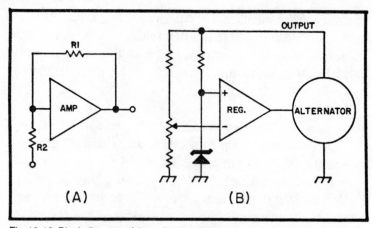

Fig. 10-16. Block diagram of the solid-state alternator system is shown in B. The system closely resembles the simple DC feeback amplified in A. In this type of amplifier, the gain approaches R1/R2. If the gain is very high, the regulator in B will not allow the preset voltage to change until the alternator is no longer capable of delivering the required current.

allow the preset voltage to change until such time that the alternator is no longer capable of delivering the required current. The 0.1 μF capacitors provide regulator stability since they apply local degeneration at a relatively high frequency. The 1N645 diode protects the output transistor when you shut off your ignition. This stops the large negative inductive surge which might conceivably achieve a value of 600V open-circuited. Any rectifier you have on hand that can handle an amp should do the job.

GAUGE NOISE ELIMINATION

After pursuing the standard treatment on on the spark plugs, the distributor, the generator and the voltage regulator of a Falcon, a clicking noise was still present that could be heard on the weaker BC stations. Apparently the noise was being radiated since it grew worse in an expressway underpass, and could be heard on a pocket transistor radio anywhere in the car. On a rough road, or by tapping the dash of the car, the clicking became a buzz. Bypass capacitors didn't seem to help.

Using the pocket radio as a pobe and pulling off wires here and there on the car, the noise was traced to the voltage regulator which supplies the fuel and temperature gauge circuit. The Falcon circuit (Fig. 10-17) is fairly standard in Ford cars and is probably typical of a number of other cars. A cheap and dirty solution is a toggle switch X to kill the circuit while listening to a weak one; however you could run out of gas or boil over without warning. Besides it is not too hard to fix if you understand the working of the simple circuit.

Under normal driving, the battery voltage varies from 12V to 15V. With the ignition switch on, the regulator contact is first closed, applying the full battery voltage to the circuit, causing the coil in the regulator to heat the bimetallic strip, which bends and eventually opens the contact. This drops the applied voltage to zero. The strip then cools; the contact is remade and the cycle repeats. The regular can be adjusted to provide an average of 5V at its output. Of course a conventional voltmeter placed at this point would flop back and forth like crazy; however, the gauges on your car are a modern version of the old hot wire meters, which depend on the heating effect of the current rather than a magnetic field. The gauges have long time constants and therefore read average values. With what appears to the gauge as a constant voltage, the circuit becomes a simple ohmmeter measuring the resistance of the rheostat in the fuel tank or in the radiator water. Each gauge has a current draw from 60 to 200 mA at 5V.

What we need is a nonclicking regulator that will take an input that varies from 12V to 15.4V and supply a constant 5V

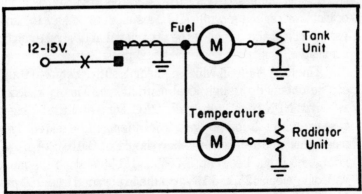

Fig. 10-17. Typical gauge hookup in Ford cars.

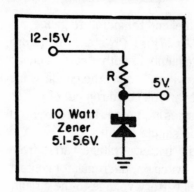

12-15 V.

R

5V.

10 Watt
Zener
5.1-5.6V.

Fig. 10-18. Easy circuit that would work if 10W zeners were readily available.

to a load that varies from 120 to 400 mA. This is a good task for semiconductors. The circuit in Fig. 10-18 would be practical if a 10W zener diode at 5.1V or 5.6V were available, or if the automaker designed his meter circuit for 6.8V or higher, where the higher current zeners are available.

The circuit in Fig. 10-19 was devised for this purpose. A 400 mW zener is used as a reference voltage controlling an NPN transistor as an emitter follower to handle the heavier current. The resistors in the circuit drop the voltages and protect the semiconductors from surges. The capacitor was found necessary for one transistor that wanted to oscillate.

The NPN transistor is something of a problem since there are not too many inexpensive ones that will handle 500 mA. The only other factor to consider is the voltage drop between the base and the emitter, and this will determine the correct zener diode voltage to use.

It was quite satisfying to install this regulator and the original car regulator with a DPDT switch to be able to compare readings and convince the doubting that the noise can be eliminated.

The RCA #40053 which nets for 1.40 in lots of 1–99 and can be obtained through local distributors. This is a silicon industrial NPN transistor in a JEDEC TO-5 package. It has a base to emitter drop averaging 0.7V among those tested. This means that the reference zener should be a 1N753A, 6.2V giving an output between 5.25V and 5.56V when the input varied between 12V and 15V and the load varied from 100 mA to 400 mA. This is a little above our 5V target but close enough

392

for the car system. A 1N752A used here gave voltages under 5V. The maximum power dissipated by the RCA 40053 is 2.2W therefore a heat sink is required. With the heat sink, the circuit ran all day long on the bench without heating too much. The 0.05 μF capacitor was found to be needed to tame some kind of an oscillation.

Construction is not critical. Just remember that the NPN transistor case can not be grounded to the car. Of course if your car has an ignition system with a positive ground, you can modify the circuit to use a PNP transistor and build it for peanuts.

ANOTHER VOLTAGE REGULATOR

A few words about the automotive battery charging system will help in understanding the operation of this circuit. An alternator's output voltage, thus its charging current, is controlled by varying the field current in the alternator. Full voltage on the field winding will give full output from the alternator, and reducing the field voltage will result in reduced output. The output capacity of any alternator is limited by its design, primarily the wire size used. In fact, loading an alternator heavily will rarely damage it, as it will put out just so much

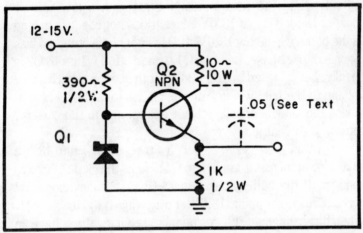

Fig. 10-19. Automotive gauge noise elimination circuit. The capacitor may be needed to stop oscillations. The transistor should be an RCA 40053. You can also use inexpensive general purpose transistors such as an ECG128, GE63, or HEP S5014.

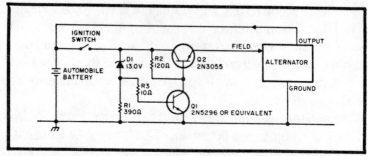

Fig. 10-20. Schematic of another type of electronic voltage regulator.

current, and beyond that its output voltage will drop, limiting the total power available.

In the usual electromechanical type regulator, a resistor is switched in series with the field winding to reduce the alternator output. When the battery voltage drops due to a heavy load, the resistor is shunted by a pair of contacts on the relay, which is voltage sensitive. Under normal conditions, with the engine running, battery voltage should be approximately 13.6V to 13.8V indicating the battery is receiving a charge.

In order to best understand the electronic circuit (Fig. 10-20), it can be looked at as a switch, either supplying no or full voltage to the field winding. Start with a low battery voltage. Diode D1, a 13.0V zener, does not conduct as long as the battery is below 13.0V. If D1 is not conducting, no bias is applied to Q1 base, keeping Q1 turned off. If Q1 is off, Q2 is on fully, being biased by R2, 120 Ω. It then applies full battery voltage to the field winding on the alternator. This of course means the alternator will put out full voltage to the battery, causing it to charge.

Now, as soon as the battery voltage increases to 13V, a small amount of bias is applied to Q1 base as the diode begins to conduct. If the battery reaches 13.6V, the zener diode will conduct fully, dropping 13.0V and leaving 0.6V to bias Q1. When this happens, Q1 turns on fully, reducing the voltage on Q2 base to very close to zero. Q2 turns off, removing all voltage from the field. Now the alternator output is reduced to zero, so the battery receives no charging current

394

In actual use, this entire process happens very rapidly and the constantly changing alternator is in effect smoothed out by the battery. Battery voltage will always be the zener diode voltage plus the drop across the base-emitter junction of Q1, about 0.6V to 0.7V. Resistor R3 limits the base current to a safe value.

The original unit was built on a large heat sink with all parts being supported by the transistor leads. This is not really necessary, as both transistors are operating as switches and consume very little power. The circuit has survived the past winter, and has always kept the battery properly charged, and at a fraction of the cost of a new regulator. It might be a handy circuit to keep in mind the next time you are left with a dead battery due to a faulty regulator, or if you are doing any experimentation with windmill power and surplus auto alternators.

LM723 CAR REGULATOR

As indicated in the schematic diagram (Fig. 10-21), this solid state automotive regulator uses a minimum of components to achieve high performance without sacrificing reliabil-

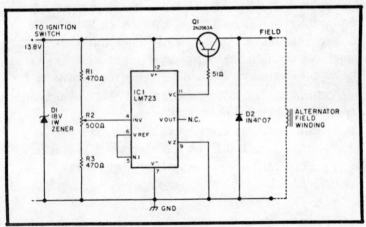

Fig. 10-21. Schematic of LM723 car regulator. D1 is an 18V 1W zener. D2 is a 1N4007 100 PIV 1A rectifier. IC1 is an LM723 14-pin DIP. Q1 is a 2N2063A 10A PNP (SK3009). R1 and R3 are 470Ω 10% resistors. R2 is a 500Ω 10-turn pot. R4 is a 51Ω ½W % resistor. Additional material: TO-3 transistor socket, 14-pin DIP socket, barrier terminal strip, TO-3 mica washer kit, PC board, iniature box, and optional relay.

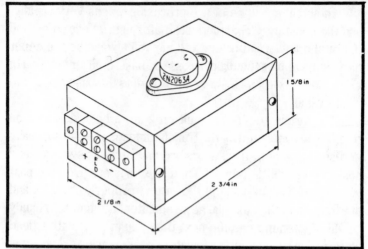

Fig. 10-22. Construction details.

ity. The heart of the unit is the LM723 precision voltage regulator IC, which is internally temperature compensated. This integrated circuit is connected as a switching type regulator to control current flow to the field of the alternator. Resistor R2 is adjusted to maintain a system voltage of 13.8V, the fully charged voltage of most standard car batteries.

If the alternator tries to produce a voltage above the set level, the LM723 turns off pass transistor Q1, thereby cutting off field excitation in the alternator. When this happens, the output voltage from the alternator begins to drop. As soon as the output level drops below 13.8V, the regulator turns the field current back on to raise the output voltage. This cycle is repeated hundreds of times a second to maintain the alternator's output voltage precisely at the set level.

External pass transistor Q1 is required to handle the large field current of most alternators (approximately 3A), since the LM723 has a maximum output current capability of 150 mA.

Construction Details

The solid state voltage regulator may be built in a small minibox (2¾ inches by 2⅛ inches by 1⅝ inches) as shown in Fig. 10-22. Transistor Q1 is mounted on top of the minibox, which is used as a heat sink. Insulate the transistor from the metal case using a TO-3 transistor socket and mica washer kit. This is necessary to prevent the transistor's case (collector) from shorting to ground.

A barrier type terminal strip (3-terminal) is used to bring the BATT, GND, and FIELD connections out. If a relay is required, you may elect to construct the unit in a larger minibox to house the relay. Also, a six-terminal barrier strip will then be required to make external connections to the relay.

In some installations, depending on the mounting location of the regulator, you may want to seal the enclosure for moisture protection; however, if the mounting location under the hood is carefully chosen, this should not be a problem.

The external pass transistor is not critical, and almost any 10A, PNP transistor will be adequate; however, plan to use only a DIP version of the LM723 and not the TO-5 version. The reason for this is that the DIP version has an internal reference zener diode (Vz) and the TO-5 version does not. The TO-5 may be used, but you will have to add an external zener reference diode. Also, the printed circuit board layout (Fig. 10-23) has been designed for the DIP version.

Installation

First, try to obtain a copy of the schematic diagram for your automotive electrical system. Most local libraries will have automotive manuals containing this type of information. You should become thoroughly familiar with this diagram before proceeding with the installation.

Referring to Fig. 10-24, determine which system best fits your own car. Four basic types of alternator systems are illustrated: Ford/Autolite, Delcotron/GM, Motorola/AMC, and the Chrysler/Plymouth system with an ammeter. With the exception of Chrysler/Plymouth, most systems will require an external relay to maintain the alternator charge indicator light function; however, if you install an external ammeter, you can eliminate the requirement of the relay. Simply connect the regulator as shown in Fig. 10-24A.

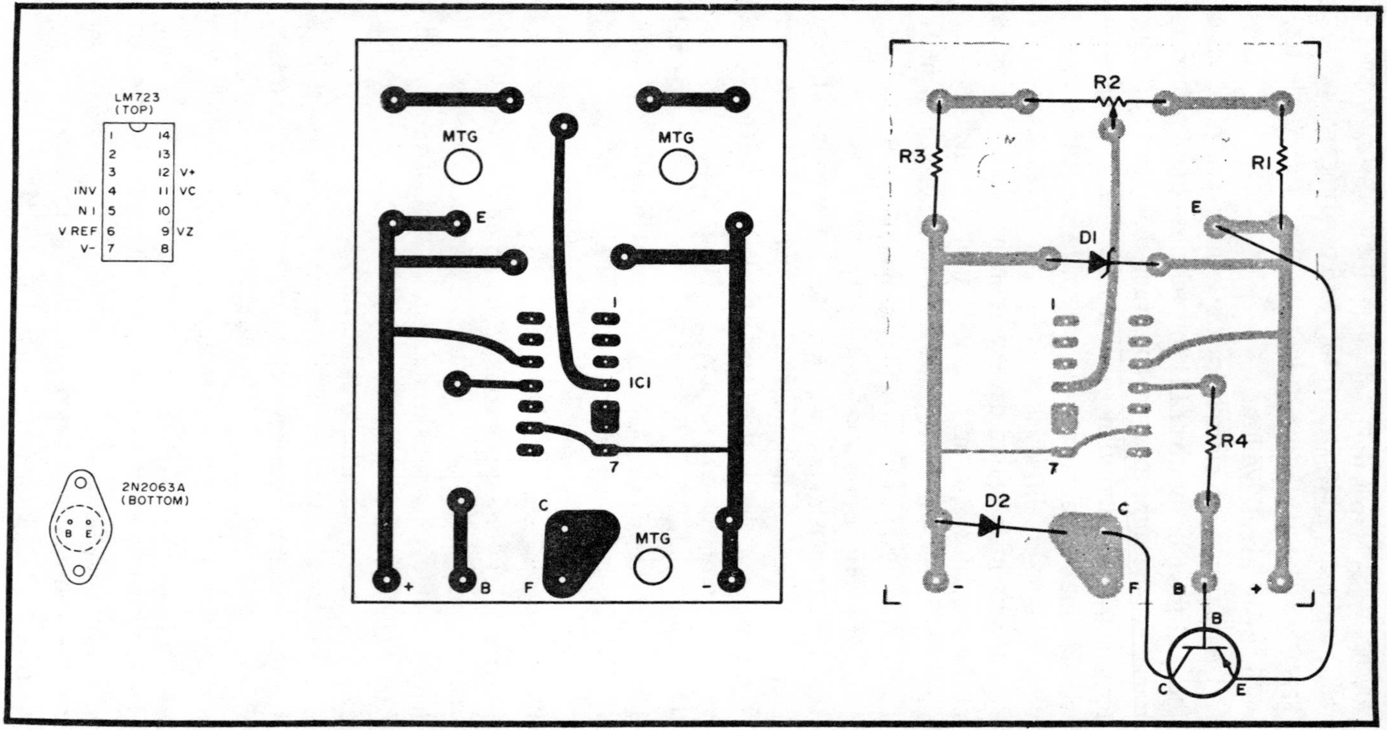

Fig. 10-23. PC board layout with basing diagrams.

The next step is to find a suitable location under the hood to mount the electronic regulator. Preferably, this location should be near the battery and away from areas subject to moisture or excessive heat.

Disconnect the old regulator and mark each of the connecting wires for future reference, and use crimp-on connectors to connect the new regulator to the system. This will maintain the integrity of the original system connections should you ever want to convert back to the original configuration. If an external relay is required, mount it in a protected space, preferably with a dust cover or within the regulator enclosure.

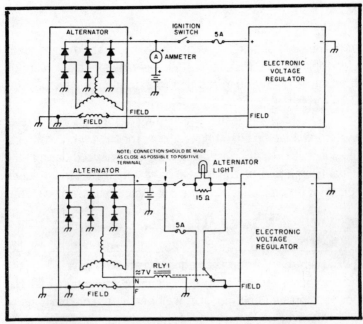

Fig. 10-24. Hookup diagrams for various makes of cars. (A) Typical hookup using an ammeter in lieu of an alternator charging light. This system does not require a relay to convert to an electronic regulator. (B) Food hookup with a charging indicator. This type of system requires an external relay to maintain the function of the charging light. For RLY1, any 6V relay with 3A SPDT contacts will do. (C) Hookup for a Delcotron (GM) electrical system with a charging indicator. This system also requires an external relay to maintain the charging light funciton. For RLY1, and 6V relay with 3A SPDT contacts will do. (D) Hookup for Motorola (AMC) electrical system with an isolation diode. An externa, relay is also required to maintain the charging light function. For RLY1, any 6V relay with 3A SPDT contacts will do.

After the unit is installed, check all wiring to insure that the system is properly connected. Before starting the engine, turn off all loads until the system voltage is properly adjusted and stable. After the engine is started, adjust the system voltage (with R2) for 13.8V at the positive terminal of the battery.

Check if the regulator is functioning properly by increasing the engine speed and adding loads to the system. The voltage should remain constant. At slow idle, with loads turned on, the voltage may drop slightly, since the alternator is not producing at its rated output. At cruising speed, however, the correct voltage should be maintained if the system is operating properly.

Conclusion

This completes the installation and checkout of your electronic voltage regulator. It should provide many years of troublefree operation in addition to extending the life of your lead-acid battery.

As a final suggestion, you may want to monitor the system voltage on a continuous basis for the first few weeks after installation. If no problems are experienced during this initial trial period, it can be safely assumed that the voltage regulator is compatible with your particular electrical system.

520V 500 mA FROM AN ALTERNATOR

In a 12V system (Fig. 10-25), alternator output will be 14V, 3-phase, AC, and the frequency will vary from 80 to 100 Hz as you speed or slow the engine. Conventional 60 Hz power supplies don't perform well under these conditions.

However, a 3-phase high-frequency supply can be easily built using mostly surplus parts, and the combination of the alternator system and the 3-phase power supply will cost only slightly more than would a conventional transistor power supply to be used with conventional auto systems.

A schematic diagram of such a unit, designed for 500V output at 500 mA (enough to power a 200W final and leave 50W to drive the exciter), is shown in Fig. 10-26.

In building the power supply, surplus 400 Hz transformers which can be obtained at a low cost can be used for the 3-phase stepup system. Proper phasing is essential for cor-

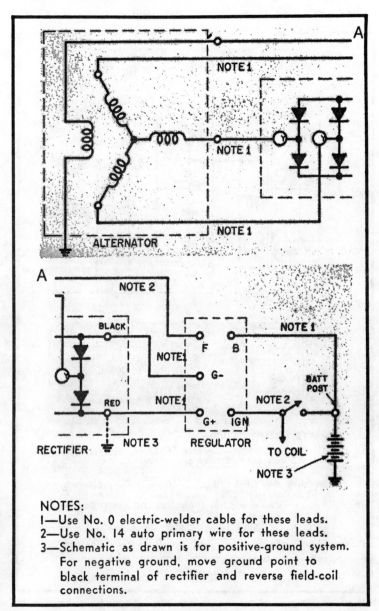

Fig. 10-25. Schematic of typical alternator installation.

NOTES:
1—Use No. 0 electric-welder cable for these leads.
2—Use No. 14 auto primary wire for these leads.
3—Schematic as drawn is for positive-ground system. For negative ground, move ground point to black terminal of rectifier and reverse field-coil connections.

rect operation. It can be assured if identical transformers are used for the three legs. Proceed as follows:

Mark one secondary terminal of each transformer *1* and the other *2*, taking care to make sure that the corresponding terminal on each transformer bears the same number. It

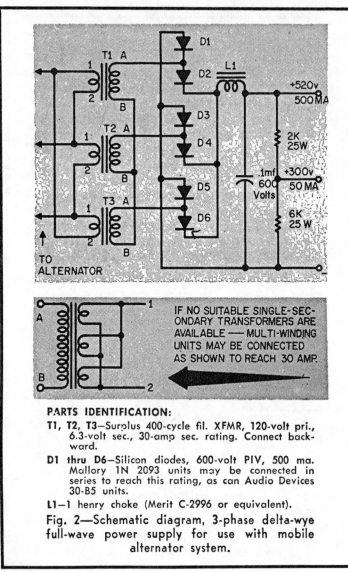

PARTS IDENTIFICATION:

T1, T2, T3—Surplus 400-cycle fil. XFMR, 120-volt pri., 6.3-volt sec., 30-amp sec. rating. Connect backward.

D1 thru D6—Silicon diodes, 600-volt PIV, 500 ma. Mallory 1N 2093 units may be connected in series to reach this rating, as can Audio Devices 30-B5 units.

L1—1 henry choke (Merit C-2996 or equivalent).

Fig. 2—Schematic diagram, 3-phase delta-wye full-wave power supply for use with mobile alternator system.

Fig. 10-26. Schematic of three-phase delta-wye full-wave power supply with parts list.

makes no difference whether the start or finish of the winding is *1*. Mark the primary terminals *A* and *B* in the same manner.

Connect the 6.3V secondaries all in series, terminal 1 of transformer I to terminal 2 of transformer II, terminal 1 of transformer II to terminal 2 of III, and terminal 1 of III to 2 of 1.

Connect all terminal Bs of the 120V primaries together. Tape the connection. Take output from the As terminal.

The delta-connected secondaries are connected to the three output terminals on the alternator. The primaries go to the 600V silicon rectifiers as shown in Fig. 10-26.

Note that very little filtering is necessary with this circuit. Input frequency varies from 80 to 100 Hz, and the delta-wye circuit has only 4% ripple at the rectifier outputs (at a frequency six times the input frequency). For many applications, ripple is low enough with no filtering. The choke and the 0.1 μF capacitor remove most of the residual 480 to 6000 Hz whistle.

If a lower voltage supply is desired, use of 12.6V transformers in place of the 6.3V units will cut output voltage in half without affecting current capability. Rectifiers then need be only 300 PIV rating. A dual-voltage supply, convenient for operating exciter, receiver, and final from the same unit, is best achieved by use of a tapped bleeder as shown in Fig. 10-26.

Chapter 11
Special Purpose AC Supplies

Most of us don't think of an AC source as being a power supply. That is because we rely on the power mains for our AC source and are not required to design circuits to convert it. Instead we design circuits to work from it.

In this short but useful chapter we deal with three types of AC supplies: high-voltage filaments for transmitter tubes, the variable-ratio autotransformer, and a variable AC supply. The high-voltage filament supply will be particularly interesting to those using linear amplifiers. The variable-ratio autotransformer information can be used by all do-it-yourselfers. See Chapter 1 about power transformer ratings and conversions for use in tube-type equipment.

LINEAR AMPLIFIER FILAMENT TRANSFORMER

A major expense when it comes to building that linear is the price of the filament transformer. Fortunately, one or two or those obsolete tube power transformers can get you over that expensive hurdle with only a little time and a few cents invested.

First weigh your transformer to determine capability. Refer to Fig. 11-1, which illustrates the relationship between weight and filament power capability. For example, if the power transformer weighs 4 pounds, it should have a filament power capability of 60W.

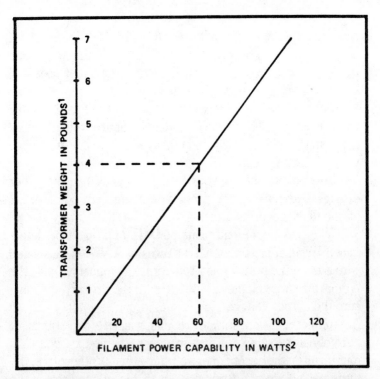

Fig. 11-1. Weight versus power capability chart for filament transformers. Weight excludes mounting fixtures. If case is heavy steel construction, remove before weighing transformer.

Next determine if the transformer is adequate for your application. Let's say you have a transformer with a recycled capability of 60W and you desire a new secondary for a 4CX1000A high power ceramic-metal tetrode. As the 4CX1000A has a maximum filament power requirement of 59.4W (6V × 9.9A), you could indeed use this transformer for your filament supply. You could also use two 30W or 40W transformers in parallel.

Once you have selected your transformer, disassemble the outer case and note the location of your primary winding in respect to the core of the transformer. The primary leads are usually color coded black. Make sure your transformer has a 110V 60 Hz primary, as some surplus transformers have odd primary voltages that operate at other than 60 Hz. Also, some transformers do have their primary near the outer core and,

405

generally, this type of transformer provides little area for your new secondary. Transformers that have the primary wound tightly around the center core provide the most area and versatility for your new secondary.

After you have located the primary, cut off and remove with a hacksaw all secondary windings that are wound *around* the primary. Work slowly and take care not to cut or damage your primary winding.

After completion of your cutting, inspect the primary for damage. Next, wrap the primary winding with one layer of plastic tape. Securely attach and insulate the primary leads.

The number of load turns per volt must next be determined. Wind approximately four turns of AWG #18 insulated wire around the primary as a temporary secondary. Apply the normal primary voltage (110V 60Hz) to the primary and measure the output voltage of your temporary secondary with an AC voltmeter. The voltmeter reading will determine the turns per volt. For example, if you measure 2V, you know it took four turns on the secondary to produce this 2V; therefore, the turns per volt is two turns per 1V. Keep in mind this is the no-load turns per volt.

After removing the temporary secondary, wind your permanent secondary. Two AWG #14 wires in parallel were used in the prototype secondaries, as this is qutie easy to wind around the primary. Two AWG #14 copper wires will be adequate for secondary current levels up to 10A, 5A per wire. The insulation on the wire should be capable of withstanding at least ten times your output voltage. Example: if the secondary output is 7V, the insulation should provide protection up to at least 70V or higher.

Always allow about 50% more wire than your turns per volt indicated, as you will have to increase the number of turns to compensate for the transformer resistance when operated under load.

A filament transformer must have the correct output voltage under load; therefore, you must load your secondary and take periodic measurements during its construction.

Let's say for example you require 6V at 10A. According to Ohm's law the load must be 0.6Ω.

$$R = \frac{E}{I} = \frac{6V}{10A}$$

Therefore you should load the secondary with a 0.6Ω resistor while measuring the output secondary voltage. Ohm's law requires the power dissipation of the resistor be at least 60W.

$$P = IE = (10A)(6V) = 60W$$

However, a much smaller wattage resistor may be used if you work rapidly and do not allow the resistor to heat up.

THE VARIABLE-RATIO AUTOTRANSFORMER

The variable-ratio autotransformer, more commonly known by trade names such as Variac, Powerstat, etc., has been with us for many years. Most of us are aware of its usefulness when used by itself to provide a variable AC voltage from the 60 Hz line. There are methods of greatly extending its usefulness which are not so commonly known, however, and some of these will be presented here.

The Variac

Perhaps we first should know what the variable-ratio autotransformer is and how it works. A note of caution is in order here. In addition to standard catalog models, each manufacturer builds an almost unbelievable number of "specials." Both types often appear later as surplus at very attractive prices. It pays to be sure of what you are ordering, and such information is free for the asking. Although the General Radio Variac will be used as an example, the principles involved will apply to variable-ratio autotransformers of all manufacture.

A schematic diagram of a typical Variac appears in Fig. 11-2. Basically, it consists of a doughnut-shaped iron core on which is a single-layer toroidal winding. The winding is tapped near each end, with the end and tap connections brought out to a terminal plate. The insualtion is ground off the winding at one

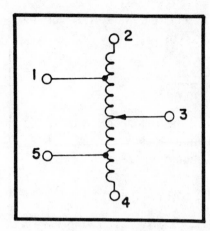

Fig. 11-2. Basic variable-ratio autotransformer.

end of the doughnut to provide a track on which a carbon brush makes contact. Electrically, the brush is connected to a terminal on the terminal board. Physically, it is attached to the shaft. The shaft may be rotated through about 320°, sufficient to allow the brush to travel from one end of the winding to the other.

In an autotransformer, the single winding acts as both a primary and the secondary. If the ends of the winding are connected across the power line as shown in Fig. 11-3, the line voltage will be equally divided by the number of turns. This results in a fraction of a volt across each turn. All the turns are utilized as the primary. The turns across which the output is taken correspond to the secondary of a conventional transformer. The number of these secondary turns, each with its volts-per-turn contribution to Vo, is selected by the brush. Thus, as the control knob is rotated, the brush contact traverses the winding, tapping off a portion of the total voltage across the winding, and the output voltage, Vo, is continuously adjustable from 0 to as large V_{LINE}. If Vo is taken between terminals 2 and 3, operation is the same except that the direction of knob rotation for increasing or decreasing Vo is reversed. In either case, Fig. 11-3 illustrates the so-called line-voltage connection.

The second basic connection is illustrated in Fig. 11-4 the over-voltage connection. By connecting one side of the input line to the end of the winding and the other side of the input line

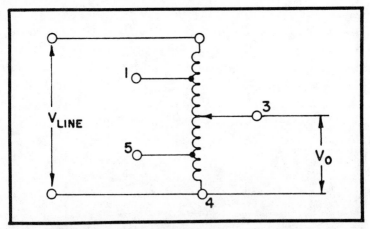

Fig. 11-3. Illustrating the line-voltage connection.

to the tap near the far end of the winding. Vo is continuously adjustable between 0 and the input line voltage as the brush moves from tap 4 to tap 1. As the brush moves from tap 1 toward the end of the winding labeled 2, additional turns and, thus, additional volts, are picked off. Variacs presently are tapped to provide a maximum output voltage 17% greater than the input voltage with this connection.

In the majority of cases it is not too difficult to determine the necessary ratings. If the Variac is to be used as described, only the line voltage and load current need to considered. The permissible load current is not the same for both connections, being slightly higher for the line-voltage connection. In either case, many times the rated current may be drawn momentarily (such as in motor starting), but brush life will be considerably reduced if the Variac is overloaded for any length of time.

Fig. 11-4. Illustrating the over-voltage connection.

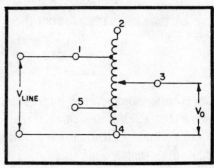

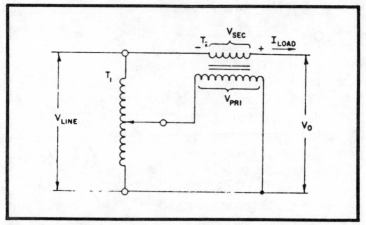

Fig. 11-5. The buck or boost scheme. Transformer T2 effectively multiplies the Variac current capacity.

Manufacturers usually publish permissible overload data in the form of time-current curves.

The Variac finds innumerable application in the two connections previously described. Another very useful circuit is shown in Fig. 11-5. Here the primary of transformer T2 is supplied with the continuously adjustable output voltage of T1. The secondary voltage of T2 is related to its primary voltage by the turns ratio.

$$V_{sec} = \frac{N_{sec}}{N_{pri}} V_{pri}$$

The secondary voltage of T2 is then connected in series between the line and the load. The polarity signs in Fig. 11-5 are used to indicate that the secondary voltage adds to the input line voltage to give the output, or load voltage.

$$V_O = V_{LINE} + V_{SEC}$$

Since from simple transformer theory the primary and secondary currents (neglecting losses) of T2 are related by

$$I_{pri} = \frac{N_{sec}}{N_{pri}} I_{sec}$$

we achieve some rather interesting results. First, if we assume that T2 is a step-down transformer of say 10-to-1 ratio and for T1 connected in the line-voltage connection, V_O may be caused to vary from V_{LINE} to $1.1\ V_{LINE}$ as the output of T1 varies from 0 to V_{LINE}. Thus, we are now able to push V_{LINE} up by as much as 10%. So what, you say. We could boost the line voltage by as much as 17% by simply using the over-voltage connection on T1 without using T2 at all. True, but the real hooker shows up in the last equation. Plugging in the numbers and turning the crank, we find that the Variac must now supply only one-tenth the load current. Thus, by using the additional transformer, we effectively mutliply the current rating of the Variac (and thus the permissible load power) by the inverse of the turns ratio of T2. If we want V_o to be lower than V_{LINE}, it is only necessary to reverse either the primary or the secondary connections of T2. Simulating line-voltate fluctuations and line-voltage correction are only two of the many applications for this technique. The increased resolution of the Variac dial greatly simplified the accurate setting of V_o. For typical application where the line voltage is approximately 120V, T1 would also be a 120V model. Commercial line-voltage regulators employing this technique typically have buck and boost range of 10%. If T1 is a 2A Variac and T2 a 120V to 12V transformer with a 20A secondary, this circuit would be capable of as much as 10% buck or boost of the line voltage applied to a 2kW load. It would therefore be quite adequate as a primary regulator in front of a 1kW transmitter.

The trouble with most power lines is that the voltage is seldom where it should be, and it is not even there for very long. At best we may consider it usually within a certain voltage range. This may mean that some days you want to boost the voltage and other days you want to buck it, in which case, reversing the phase of the secondary voltage of T2 (in Fig. 11-5) could get pretty tiresome. There is a very simple way out as illustrated in Fig. 11-6. Here we insert a special tap in the electrical center of the Variac winding. Now the voltage applied to the primary of T2 is 0 when the Variac knob is in the center of its travel. Rotating the knob in either direction from this point will increase the voltage applied to the primary. However, rotating the knob in one direction will cause a secondary buck voltage, while rotating it the other way will cause a secondary boost voltage. Thus, we get our buck or boost voltage without switching, merely by adding a tap to the center of the Variac winding. Of course, there is a catch to it,

411

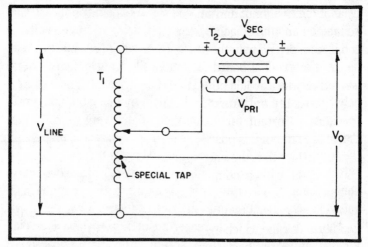

Fig. 11-6. Method of obtaining both buck and boost operation without switching.

and we still don't get something for nothing. Notice that now the maximum voltage we can apply to the primary of T2 is only half the line voltage. If we want to utilize the maximum power capability of T2, we require a transformer with a 60V primary winding (when V_{LINE} is 120V). Granted, these are a little hard to come by. Such animals are in existence, and you might be lucky enough to come across one.

Otherwise, there are two possibilities if you think the work is worth it. One is to start from scratch and build your own; the other possibility is to unwind a transformer, such as a large TV power transformer, carefully noting the number of turns on each winding. The primary and secondaries may then be rewound to get exactly the voltages and ratios you want. One little trick here is to effectively break the original primary in half and then connect the two halves in parallel. It is very important that the number of turns on each half be identical, so you still have to unwind the primary and count the turns. The winding is then broken when rewinding the primary.

Tapping the Variac is a very simple operation. Though not essential, the exact center turn of the winding may be located if you're a purist. The center of the winding will be halfway around the doughnut from the terminal board. The turn selected should be lifted carefully away from the outer

412

edge of the doughnut and a piece of insulating material slipped between it and the adjacent turns. The wire should be cleaned and tinned. A lead may then be soldered to it and then brought out at the terminal board.

Of course it would be very nice to have both the buck and boost capability with neither a specially tapped Variac nor a special transformer for T2. This sounds like something for nothing again, but it is possible as shown in Fig. 11-7. Here we utilize the fact that the line coming into most homes is a split 230V circuit with a common center. If T1 is now a 230V Variac, we may use the center tap on the input line where we formerly required a center tap on T1. Additionally, since the maximum voltage which may be applied to the primary of T2 is again 115V, we are back to the garden-variety transformer for the buck-boost voltage.

Wired as shown, we do get both buck and boost operation without switching transformer leads. The only problem now is to watch out for the ratings of T1 and the availability of the split 230V line. In using such a system, remember there are such things as building codes. As to the rating of T1 and T2, there should be no problem if you just revert back to our simple equations.

A Versatile Distribution System

One fairly useful combination of techniques previously described is illustrated in Fig. 11-8. In this circuit, T1 is a

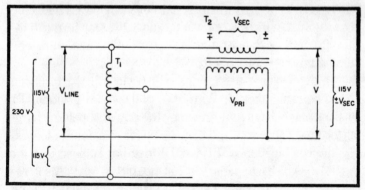

Fig. 11-7. Buck and boost scheme wherein no tap is required on T1, and T2 may have a 120V primary.

413

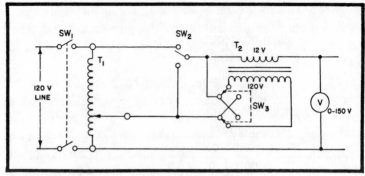

Fig. 11-8. A versatile power distribution system.

standard 120V Variac and T2 is a 12V filavent transformer. With SW2 in the position shown, V_o, is adjustable from 0 to V_{LINE}. With SW2 in the upper position, V_o is adjustable from V_{LINE} to 10% above or below V_{LINE}, depending on the setting of SW3. This sytem is quite useful for bench work, and if the loads to be encountered are relatively small, either the line voltage or the over-voltage connection may be used for T1.

Another particularly useful application for this system is for primary power control for a single sideband transmitter. The lower position of SW2 permits the DC to be brought up slowly to charge the normally high value of filter capacitance in the power supply. The upper position of SW2 then allows the line voltage to be maintained at the desired fixed level. In this application it might be desirable to be able to switch the voltmeter between the input and output line to determine which position of SW3 will be required. Switch SW3 should be set with SW2 in the lower position and T1 set for zero output. This prevents T2 operating as a current transformer and burning up the contacts of SW3.

In determining the ratings of T1 and T2, the following should be remembered. With SW2 in the lower position, T1 supplies only the current drawn by the power supply in charging the filter capacitors and to make up for transformer losses. It is the combination of T1 and T2 together as described for Fig. 11-5 which supplies the full operating load. Also, remember that the contacts of SW1 and SW2 must now be capable of carrying the full output-load current.

VARIABLE AC SUPPLY

In the realm of electronic experimentation, it becomes necessary to use a variable voltage source as a means of precision voltage control.

The variable AC circuit outlined in Fig. 11-9 provides a full range of voltage control for the primary of any 120V 60Hz transformer requiring less than 10A of primary current.

This circuit can be constructed in a small 3- by 5-inch utility box, or mounted in the front panel of the controlled

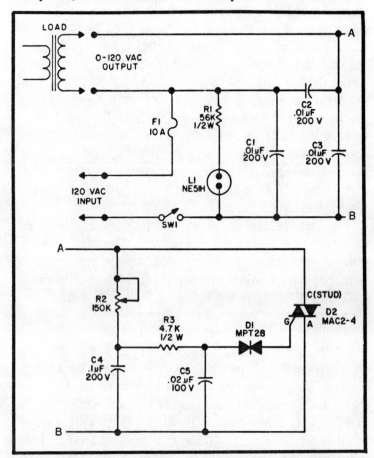

Fig. 11-9. Schematic of variable AC supply. C1 through C3 are 0.01 mF 200V capacitors. C4 is a 0.1 μF 200V capacitor. C5 is a 0.02 μF 100 V capacitor. D1 is a MPT28 (ECG6407). D2 is a MAC2-4 200 V traic (ECG5663). F1 is a 10A fuse, L1 is a NE51H neon lamp. R1 is a 56K ½W 10% resistor. R2 is a 150 K linear taper ½W pot. R3 is a 4.7K ½W 10% resistor. SW1 is a SPST 10A switch.

415

source. The only modification to the equipment is a single ⅜-inch mounting hole for the 150K pot. If an outboard system is more versatile, then a line cord and socket will be needed for ease of connection.

Construction

Choose either of the two above mentioned mounting arrangements and begin by mounting the MAC2-4 triac to a good heat sink surface. Since this is a stud mounted device, a single ¼-inch hole will be needed. Apply silicon grease or heat sink compound to both insulating washers and the base of the device before mounting. Check for electrical shorts between the stud and chassis after mounting the triac, for the stud carries full line potential. When connecting the MPT-28 trigger diode, be certain to heat sink the leads before soldering. The leads should be left full length for heat dissipation. Both leads of the MPT-28 are identical and no polarity need be observed. Although the anode and cathode of the MAC2-4 are identical, the gate is the shorter of the two terminals protruding from the top of the device, and it must be connected to the trigger diode MPT-28.

Addition of the on-off switch, fuse, neon lamp, and capacitors (C1, C2, C3) are all optional. The capacitors are installed to eliminate the RF or noise generated by the system. Experimentation with the requirements for these capacitors is left up to the individual.

Applications

The control of many types of AC loads is within the capabilities of this simple circuit. One example is the addition of this AC supply to the circuit shown in Fig. 11-10. A battery charger or high current bench supply that is variable from 0V to 30V DC is very useful on the bench for powering bread board circuits. Figure 11-10 shows a straightforward power supply using a full wave bridge if needed. A well stocked junk box should yield most of the parts required for the supply.

Other applications include a variable supply for hobby use such as electric trains or slot cars, plating of metals, and light

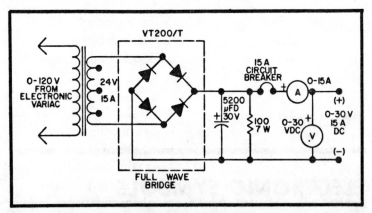

Fig. 11-10. DC battery eliminator/charger/power supply. Meter A is a 0 to 15A ammeter. Meter V is a 0 to 30V voltmeter. The transformer is a 24V 15A filament type.

control. It can be used to replace that worn out adjustable transformer in the plate supply of your high power rig. This circuit does the job of a large variable transformer costing more than three times as much.

Appendix

ELECTRONIC SYMBOLS

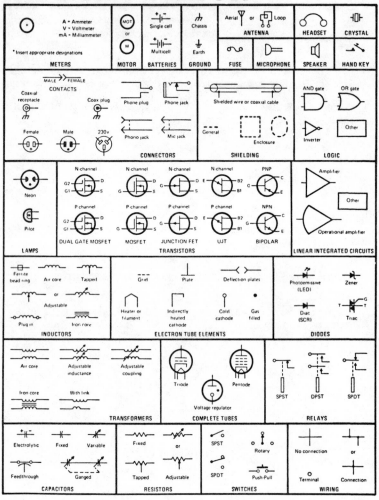

Index